ALSO BY MICHAEL SHERMER

The Moral Arc: How Science and Reason Lead Humanity Toward Truth, Justice, and Freedom

The Believing Brain: From Ghosts and Gods to Conspiracies and Politics—How We Construct Beliefs and Reinforce Them as Truths

The Mind of the Market: Sharing Apes, Trading Humans, & Other Tales of Evolutionary Economics

Why Darwin Matters: Evolution and the Case Against Intelligent Design

Science Friction: Where the Known Meets the Unknown

The Science of Good and Evil: Why People Cheat and Lie, Share and Care, and Follow the Golden Rule

The Borderlands of Science: Where Sense Meets Nonsense

Denying History: Who Says the Holocaust Never Happened and Why Do They Say It?

In Darwin's Shadow: The Life and Science of Alfred Russel Wallace

How We Believe: Science, Skepticism, and the Search for God

Why People Believe Weird Things: Pseudoscience, Superstitions, and Other Confusions of Our Time

The Skeptic Encyclopedia of Pseudoscience (General Editor)

The Secrets of Mental Math (with Arthur Benjamin)

SKEPTIC

SKEPTIC

VIEWING THE WORLD WITH A RATIONAL EYE

MICHAEL SHERMER

HENRY HOLT AND COMPANY
NEW YORK

Henry Holt and Company, LLC
Publishers since 1866
175 Fifth Avenue
New York, New York 10010
www.henryholt.com

Distributed in Canada by Raincoast Book Distribution Limited.

The essays in this book originally appeared, in slightly different form, in *Scientific American* magazine.

Library of Congress Cataloging-in-Publication Data

Shermer, Michael.
Skeptic: Viewing the World with a Rational Eye / Michael Shermer.
pages cm
Summary: Seventy-five of author's contributions to Scientific American magazine.
ISBN 978-1-62779-138-0 (hardcover)—ISBN 978-1-62779-139-7 (ebook) 1. Science—Philosophy. 2. Skepticism. I. Scientific American. II. Title.
Q175.S53438 2005
501—dc23 2015003697

Henry Holt books are available for special promotions and premiums. For details contact: Director, Special Markets.

First Edition 2016

Designed by Kelly S. Too

Printed in the United States of America
1 3 5 7 9 10 8 6 4 2

To my sister Tina

CONTENTS

II. SKEPTICISM

III. PSEUDOSCIENCE AND QUACKERY

IV. THE PARANORMAL AND THE SUPERNATURAL

V. ALIENS AND UFOS

VI. BORDERLANDS SCIENCE AND ALTERNATIVE MEDICINE

SKEPTIC

Introduction

Viewing the World with a Rational Eye

Ever since the early 1980s, when I discovered the elegant and entertaining essays by the late Harvard evolutionary biologist and paleontologist Stephen Jay Gould—initially through his early essay collections in books (*Ever Since Darwin* and *The Panda's Thumb*) and subsequently in monthly form as they rolled off the presses for *Natural History* magazine—I have maintained a deep passion to write science for general audiences. Not "popular science" writing per se, but more along the lines of what Gould strove for in his essays: deeper truths within scientific discoveries. As he wrote in the preface to a later collection of essays, *The Lying Stones of Marrakech*: "I have tried, as these essays developed over the years, to expand my humanistic 'take' upon science from a simple practical device . . . into a genuine emulsifier that might fuse the literary essay and the popular scientific article into something distinctive, something that might transcend our parochial disciplinary divisions for the benefit of both domains (science, because honorable personal expression by competent writers can't ever hurt; and composition, because the thrill of nature's factuality should not be excluded from the realm of our literary efforts)." Transdisciplinary doesn't begin to describe the breadth and depth of Gould's oeuvre, and it has been a mark at which I have aimed in my own writing.

In a 2002 paper published in the journal *Social Studies of Science* I presented the results of a content analysis I conducted of all three hundred essays Gould penned in his twenty-five-year monthly streak, revealing five deeper themes that appeared in them: Data-Theory, Time's Arrow-Time's Cycle, Adaptationism-Nonadaptationism, Punctuationism-Gradualism, and Contingency-Necessity. The first theme, on how data and theory interact, interested me the most, inspired as I was by a quote Gould occasionally employed from his hero Charles Darwin: "How odd it is that anyone should not see that all observation must be for or against some view if it is to be of any service!" The context for that quote, as I explicate in the first essay in this volume (in what I call "Darwin's Dictum"), was that Darwin was challenged by his critics to just put his data forward and not to bother theorizing too much. But as the founder of evolutionary theory knew, the facts never just speak for themselves; they are always viewed through the lenses of theory. The two—observations and views, data and theory—are the conjoined twins of science.

That theme—the interplay of data and theory—is the central and unifying schema in all the essays in this volume. Gould completed his streak of three hundred consecutive monthly essays in January 2001. My first essay in *Scientific American* was in April of that year, but with a three-month publication lead time I actually penned the first one that January, close enough for an intellectual transition (if only in my mind). The seventy-five essays in this volume are from the first six and a quarter years of my own streak that—good health and fortune willing—will reach three hundred in April 2026, with several more essay volumes to come. I have grouped them into ten sections by topic within the larger data-theory schema.

I. *Science.* In this section I set the tone of the entire series with Darwin's Dictum in the first essay, giving examples of why data and theory are necessarily bound together, then move through essays whose themes related to general scientific principles and debates, such as what scientists should say when they're wrong ("I was wrong" is a good start) and what it means to be wronger than wrong, scientism and why people look up to scientific superstars like Stephen Hawking, how to tell the difference between the ideas of a cutting-edge scientist versus a cutting-edge amateur, how to communicate science in both words and pictures, and the nature of scientific replication.

II. *Skepticism.* These essays range from classic skeptical debunkings (from the "faked" moon landing to 9/11 conspiracy theories) to the delicate balance between orthodoxy and heresy in science and when we should be skeptical of a heretical idea, and the fine art of baloney detection.

III. *Pseudoscience and Quackery.* There is also a difference between heresy and crazy, and the essays in this section explore why smart people fall for the latter, why being too open-minded is not always a good idea, why scams and cons work, and why bad ideas are dangerous to the point where they can kill.

IV. *The Paranormal and the Supernatural.* The essays in this section are about nothing. That is to say, there is no such thing as the paranormal and the supernatural; there is just the normal, the natural, and mysteries we have yet to explain. Talking to the dead, ESP and PSI, Bible codes, random electronic signals that sound like voices, illusory patterns in Beatles' albums, and what happens when a skeptic goes to the New Age capital of the world.

V. *Aliens and UFOs.* This is one of my favorite of all subjects in the skeptical pantheon of things to investigate, because there are really two different questions: Are aliens out there? and Have aliens come here? The answers to these two questions are "probably" and "probably not." These essays consider what it means to search for extraterrestrial intelligence, and how we would know when we made contact, a good reason why we have not yet heard from ET, what time travel means for the search, and what it feels like to be abducted by aliens.

VI. *Borderlands Science and Alternative Medicine.* For my money the most interesting ideas are not those that are obviously right or wrong, but those that, if they are true, would be revolutionary. Nanotechnology and what it means for immortality, cryonics, cloning, and cures for the common cold, and other remedies that promise us the world but rarely deliver.

VII. *Psychology and the Brain.* I am by training a psychologist, so I am always interested in how our brains work, especially how we can so easily be fooled and mistaken in believing things that are not true. The essays in this section explore the many facets of human psychology as they relate to our beliefs about the world, especially how and when our intuitions serve us well or lead us astray.

VIII. *Human Nature.* Fundamental to who we are and why we think and act as we do is our evolved biological nature, and the essays here consider a number of highly controversial scientific theories in this regard. For example, are we noble or ignoble savages? Are we biologically predisposed toward making love and war? Can we unweave the heart to understand love and attachment? And what is happiness, anyway, and can science measure it?

IX. *Evolution and Creationism.* Ever since I was in college this topic has periodically erupted on the political and cultural landscape as scientists hope creationism goes away while creationists continue to evolve new strategies to wedge their ideas into the minds of the public, including and especially students. These essays dip into that controversy from a number of different points, ranging from science to politics.

X. *Science, Religion, Miracles, and God.* Outside of the specific case of creationism, which focuses on evolution as a perceived threat to a narrow sect of religious belief, there is arguably no more contentious topic in all of science today than how it relates to religion, miracles, and God (most notably the latter's existence—or not). It seems to be a perennial topic for scientists and philosophers of all stripes to ring in on, with new books landing on my desk almost weekly since I began this column a decade and a half ago. That's a lot of ink spilled to solve a problem that may be insoluble. Is it? The essays in this final section consider that question, and others, in what is essentially a guide to how to think about science and religion, even while offering my own often strongly worded opinions.

Writing monthly essays for *Scientific American* has been one of the most consistent joys in my life. I look forward each month to exploring a new theme or topic, chosen based on a number of criteria partially shaped by the editorial policies of the magazine, which has been in print for more than a century and a half (so they know what they're doing). These include new and newsworthy findings, discoveries, experiments, surveys, articles, and books in science that people care about and—fitting with my data-theory schema—in each essay I try to connect them to some deeper theoretical idea with some relevance for society and culture.

My pathway in this series has been navigated around many rocky shoals by the remarkable editors at *Scientific American*, including most notably John Rennie, Mariette DiChristina, and Fred Guterl, all of whom have struck the perfect balance of editorial hand, improving my prose without rewriting to the point of unrecognizability. And I would especially like to thank the fact checkers at *Scientific American*, most notably Aaron Shattuck, who has caught some potentially embarrassing whoppers before they went to press. In cycling we have a saying: there are only two types of cyclists—those who have crashed and those who are going to crash. Well, in literature there are only two types of writers—those who need editing and those who are going to need editing. The *Skeptic* column—and by extension this book—would not be possible without such supreme editing. I thank them, and the readers of *Scientific American* who have consistently supported my column, for this great honor.

A final note on length and content: To the original essays of this collected series I have, where appropriate, added updates that include corrections and addendums of new information that have come to light since the original publication dates. As well, I have included the slightly longer versions of the essays as I originally penned them for *Scientific American*; that is, my original published essay length was one page with an illustration, which delimits the word count to about seven hundred. But I usually wrote about eight hundred to a thousand words and then started cutting, a painful process that I endured by promising myself that one day the full-length versions would come to light (the "director's cut," as it were). It's not that there was anything wrong with the original seven-hundred-word pieces, only that they breathe a little freer when I'm allowed to flesh out a thought or explain an idea with a few additional sentences as I attempt to view the world with a rational eye, the reading line for the *Skeptic* column.

I

SCIENCE

1

Colorful Pebbles and Darwin's Dictum

Science is an exquisite blend of data and theory

In 1861, less than two years after the publication of Charles Darwin's *On the Origin of Species*, in a session before the British Association for the Advancement of Science a critic claimed that Darwin's book was too theoretical and that he should have just "put his facts before us and let them rest." In a letter to his friend Henry Fawcett, who was in attendance in his defense, Darwin explained the proper relationship between facts and theory:

> About thirty years ago there was much talk that geologists ought only to observe and not theorize; and I well remember someone saying that at this rate a man might as well go into a gravel-pit and count the pebbles and describe the colours. How odd it is that anyone should not see that all observation must be for or against some view if it is to be of any service!

There are few thinkers in Western history with more profound insights into nature than Charles Darwin, but for my money this is one of the deepest single statements ever made on the nature of science itself, particularly in the understated denouement. If scientific observations are to be of any use, they must be tested against a theory, hypothesis, or model. The facts never just speak for themselves, but must be interpreted through the colored lenses of ideas—percepts need concepts.

When Louis and Mary Leakey went to Africa in search of our hominid ancestors, they did so not based on any existing data, but on Darwin's theory of human descent and his argument that because we are so obviously closely related to the great apes, and the great apes live in Africa, it is here that the fossil remains of our forebears would most likely be found. In other words, the Leakeys went to Africa because of a concept, not a percept. The data followed and confirmed this theory, the very opposite of the way we usually think of science working.

If there is to be an underlying theme in this column—a substrate beneath the surface topography (to continue the geological metaphor)—it is that science is an exquisite blend of data and theory, facts and hypotheses, observations and views. If we think of science as a fluid and dynamic way of thinking instead of a staid and dogmatic body of knowledge, it is clear that the data/theory stratum runs throughout the archaeology of human knowledge and is an inexorable part of the scientific process. We can no more expunge ourselves of biases and preferences than we can find a truly objective Archimedean point—a God's eye view—of the human condition. We are, after all, humans, not gods.

In the first half of the twentieth century philosophers and historians of science (mostly professional scientists doing philosophy and history on the side) presented science as a progressive march toward a complete understanding of Reality—an asymptotic curve to Truth—with each participant adding a few bricks to the edifice of Knowledge. It was only a matter of time before physics (and eventually even the social sciences) would be rounding out their equations to the sixth decimal place. In the second half of the twentieth century professional philosophers and historians took over the profession and, swept up in a paroxysm of postmodern deconstruction, proffered a view of science as a relativistic game played by European white males in a reductionistic frenzy of hermeneutical hegemony, hell bent on suppressing the masses beneath the thumb of dialectical scientism and technocracy. (Yes, some of them actually talk like that, and one really did call Newton's *Principia* a "rape manual.")

Thankfully, intellectual trends, like social movements, have a tendency to push both ends to the middle, and these two extremist views of science are now largely passé. Physics is nowhere near that noble dream of explaining everything to six decimal places, and as for the social sciences, as a friend from New Jersey says, "fuhgeddaboudit." Yet there is progress in science, and some views really are superior to others, regardless of the color, gender, or country of origin of the scientist holding that view. Despite the fact that scientific data are "theory laden," as philosophers like to say, science is truly different from art, music, religion, and other forms of human expression because it has a self-correcting mechanism built into it. If you don't catch the flaws in your theory, the slant in your bias, or the distortion in your preferences, someone else will. Think of N-Rays and E-Rays, polywater and the polygraph. The history of science is littered with the debris of downed theories.

In future columns we will be exploring these borderlands of science where theory and data intersect. As we do so, let us continue to bear in mind what I call Darwin's Dictum: *all observation must be for or against some view if it is to be of any service.*

2

Contrasts and Continuities

Eastern and Western science are put to political uses in both cultures

In the fifth century BC, Siddhartha Gautama—better known to us as the Buddha—extolled the virtues of enlightenment through a middle path between extremes:

> Avoiding the two extremes the Buddha has gained the enlightenment of the Middle Path, which produces insight and knowledge, and tends to calm, higher knowledge, enlightenment, Nirvana. This is the noble Eightfold Way: namely, right view, right intention, right speech, right action, right livelihood, right effort, right mindfulness, right concentration.

Twenty-five centuries later the physicist Murray Gell-Mann constructed a subatomic model he playfully called the Eightfold Way, because it consisted of eight particles with eight possible rotations. It was a joke, he told a Caltech audience in a lecture, "Quantum Mechanics and Flapdoodle," that I attended, referring to the New Age fiddle-faddle about his theory presented in books whose authors didn't get the humor and thus constructed elaborate and imaginary links between Eastern mysticism and Western science. Such comparisons do tug at one's inner sense that the continuities between Eastern and Western worldviews should reflect some deeper structure, but is it really possible (in an analogy with the "uncertainty principle" in quantum mechanics) that the orbit of Mars, like the orbit of an electron, is scattered randomly around the sun until someone observes it, at which point the wave function collapses and it appears in one spot? No. Quantum effects wash out at large scales. Microcosms do not correspond to macrocosms. And the vague similarities between Eastern and Western models are the result of the fact that there are only so many variations on explanations of the world and, by chance, some are bound to resemble each other.

I was struck by such East-West contrasts and continuities on several levels during a recent trip to Beijing for the International Conference on Science Communication (which for much of China means such scientific basics as birth control, depicted in the natural history museum so graphically that captions, which I could not read, were not necessary). Held at the Chinese Association for Science and Technology in a sleek modern downtown high-rise, it was ironic that the overhead, slide, video, and PowerPoint projectors routinely broke down. Throughout the city bicycles far outnumber cars, buses, and taxis, while businessmen and -women, before cycling to their jobs in this rapidly developing technological society, flock to city parks to perform tai chi, the ancient art of adjusting one's spiritual energy.

Even at tourist attractions such contrasts abound. A tour of the Great Hall of the People at Tian'anmen Square (communism at its worst) forces visitors to exit through a basement filled with kitsch and crafts of the tackiest sort (capitalism at its worst). The Museum of Science and Technology featured an old and faded IMAX film (*The Dream Is Alive*) projected onto a water-stained, chipped-tiled ceiling; and a fabulously clever pneumatic bed of nails would have demonstrated the harmless distribution of mass over many points . . . if only it worked. Even in the Forbidden City—where emperors and empresses, concubines and eunuchs, palanquins and peons roamed for five centuries—there could not have been a more striking contraposition in the only store I found in the palace interior: a Starbucks! Of course I had to imbibe.

For my yuan (eight to a dollar), however, the finest example of contrast and continuity was the Ancient Beijing Observatory, built in 1442 for the sixth Ming dynasty emperor, Zhengtong. Located on the main east-west corridor of the city (itself laid out according to celestial coordinates) on the roof of what was once a tallish building, this observatory contains a sextant, theodolite, quadrant, altazimuth, several armillae, and a celestial globe, allowing Chinese astronomers to track the motion of planetary bodies, record eclipses and comets, and mark the location of the Milky Way galaxy and the constellations. This was the Keck Observatory of its age, measuring, for example, the length of the solar year at 365.2425 days, off by only 26 seconds. Its beautifully crafted bronze instruments stand in stark contrast to the steel girders and scaffolding that abound in high-rises going up faster than McDonald's.

A closer examination of these astronomical instruments, however, reveals interesting contrasts through several East-West continuities. The rings of the armillary sphere, for example, are divided into 360 degrees—a European tradition adopted from Mesopotamian geometry—instead of 365.25 daily segments found in pure Chinese instruments. The celestial globe presents the Milky Way galaxy in dimpled metal cutting a swath across the globe, while rough cut metallica stars explode from the surface marking a most familiar constellation, Orion, with the three unmistakable belt stars pointing to Sirius, the brightest star in the sky, the large upper corner star marking the red giant Betelgeuse, and its diagonal opposite representing Rigel. I could even find the Orion Nebula represented as the middle of three small stars just below the belt.

But then I noticed that something was amiss in the globe. Orion is backward. Betelgeuse should be in the upper left corner of the constellation, not the right, and Sirius should be to the left of the belt stars. Then I realized that the sky is inside out. According to archaeoastronomer Ed Krupp, all celestial globes are constructed from "the transcendental eye's view" of an outsider looking in. It turns out that this celestial globe (along with the rest of the instruments) was built in 1673 (during the Qing Dynasty) by a Belgian Jesuit named Ferdinand Verbiest (for measuring the altitudes and azimuths of celestial bodies), and, in Krupp's words, "blends a clearly Western pedigree with representations of traditional Chinese constellations."

The deeper purpose of this observatory reveals one final contrast and continuity of East and West, old and new. Such celestial precision was not needed for any scientific reasons in these early centuries. Rather, as Krupp explains in his insightful book on the politics of astronomy, *Skywatchers, Shamans, and Kings*, "as a truthful mirror of nature, astronomy was official business, a tool in the service of the social and political agenda of the state." Astronomical accuracy was "celestial certification of imperial power." The emperor was supposed to be the son of the celestial god Shangdi, and thus state-sponsored astronomy validated his link to the highest order and solidified the connection he represented between Heaven and Earth, sacred and profane, macrocosm and microcosm. China was the "middle land," the center of the world, with the Tian'anmen "Gate of Heavenly Peace" leading into the Forbidden City (itself aligned by the cardinal directions), followed by the "Hall of Supreme Harmony" due north on the cosmic axis, where the emperor held audiences to announce the calendar, New Year, and winter solstice.

In parallel fashion, during the conference on science communication a delegation of representatives of both Chinese and American scientific organizations had an audience with the vice premier of the State Department, which amounted to little more than a bureaucratic formality of tea and polite dialogue. As we sat patiently listening to the translation I was struck by the symbolism of the act: because science is now the royal road to reality, and communicating science is the connection between the sacred and the profane in a secular scientific society, it must be part of official state business—a certification of political power—be it monarchical Europe and imperial China, or capitalist America and Communist China. While some East-West comparisons, such as the Eightfold Way of physics, are chimerical, others are not, particularly those of a political nature, for as another ancient philosopher, this one from the West, observed, "Man is by nature a political animal."

3

I Was Wrong

Those three words often separate the scientific pros from the posers

My friend James Randi speculates—with only partial facetiousness—that when one receives a PhD a chemical is secreted from the diploma parchment that enters the brain and prevents the recipient from ever again saying "I don't know" and "I was wrong." I don't know if this happens to all PhDs, but as one counterexample I hereby confess that in my column on Chinese science in the July 2001 issue of *Scientific American* I was wrong in my conversion of Chinese yuan as eighty to the dollar (it is eight, as is noted in chapter 2 of this book). Even though I had just visited Beijing and had a Chinese colleague read the essay, the mistake still slipped through. Fortunately there was no dearth of readers who called my attention to it.

More serious was a statement I made in the June 2001 issue about the Fox program claiming that the moon landing was hoaxed. I said that the reason there was no rocket exhaust from the lunar lander is that there is no atmosphere on the moon. I was partially wrong. The lack of an atmosphere plays a minor role; the main reason is that the LEM engine used hypergolic propellants (dinitrogen tetroxide and Aerozine 50) that ignite upon contact and burn very cleanly (compare the space shuttle's nearly invisible rocket flame to the very visible solid rocket boosters' exhaust plume). Again, readers were kind enough to provide constructive criticism.

This process of critical feedback is the lifeblood of science, as is the willingness (however begrudgingly) to say "I was wrong" in the face of overwhelming evidence to the contrary. It does not matter who you are or how important you think your idea is, if it is contradicted by the evidence it is wrong. (Of course, if your name is Einstein, Feynman, or Pauling you may initially receive a more favorable hearing, but as Hollywood pundits say about the extensive studio promotion of a film, that will only buy you a week—after that it stands or falls on its own merits.) By contrast, pseudoscientists typically eschew the peer-review process to avoid the inevitable critical commentary that is an integral part of healthy science. Consider, for example, Immanuel Velikovsky's controversial theory about planetary collisions first proffered in 1950. Velikovsky was not a scientist and he rejected the peer review process after submitting a paper to the prestigious journal *Science*. "My [paper] was returned for rewriting after one or two reviewers took issue with my statement that the lower atmosphere of Venus is oxidizing. I had an easy answer to make . . . but I grew tired of the prospect of negotiating and rewriting."

Nearly a quarter of a century later, after a special session devoted to his theory was organized by Carl Sagan at the 1974 AAAS meeting, Velikovsky boasted that "my *Worlds in Collision*, as well as *Earth in Upheaval*, do not require any revisions, whereas all books on terrestrial and celestial science of 1950 need complete rewriting . . . and nobody can change a single sentence in my books." The unwillingness to submit to peer review and the inability to admit error are antitheses of good science.

A splendid example of honorable science can be found in the May 11, 2001, issue of *Science*, in the report "African Origin of Modern Humans in East Asia." A team of Chinese and American geneticists sampled 12,127 men from 163 Asian and Oceanic populations, tracking three genetic markers on the Y chromosome. What they discovered was that every one of their subjects carried a mutation at one of these three sites that can be traced back to a single African population some thirty-five thousand to eighty-nine thousand years ago. Their modestly worded conclusion that "the data do not support even a minimal in situ hominid contribution in the origin of anatomically modern humans in East Asia" is, in fact, a major victory for the "out of Africa" hypothesis that suggests all modern people can trace their heritage to Africa. It is also a significant blow against the "Multiregional" hypothesis that argues modern human populations had multiple origins dating back many hundreds of thousands of years. The finding corroborates earlier mitochondrial DNA (mtDNA) studies, the fossil record, and a remarkable discovery that Neanderthal DNA shows no signs of interbreeding with humans living contemporaneously.

One of the chief defenders of Multiregionalism, anthropologist Vince Sarich from UC Berkeley, is well known for his vigorous and energetic defense of his beliefs and theories. (I know Vince and can attest to the fact that he is a tenacious fighter for his views.) Yet when this self-proclaimed "dedicated Multiregionalist" saw the new data, he confessed: "I have undergone a conversion—a sort of epiphany. There are no old Y chromosome lineages [in living humans]. There are no old mtDNA lineages. Period. It was a total replacement." In other words, in a statement that takes great intellectual courage to make, Sarich said "I was wrong."

Whether Sarich is wrong about his conversion remains to be seen, as additional studies confirm or deny the findings (and one staunch defender of the beleaguered hypothesis told me that Vince was never really a Multiregionalist and that this study disproves nothing). The point is that creationists and social critics who decry science as dogmatic obedience to authority and an old-boys network of closed-minded fogies are simply wrong. Science is in constant flux, theories are battered about by the always shifting winds of evidence, and scientists really do change their minds.

4

The Shamans of Scientism

On the occasion of Stephen W. Hawking's sixtieth trip around the sun, we consider a social phenomenon that reveals something deep about human nature

In 1998 God appeared at Caltech.

More precisely, the scientific equivalent of the deity in the form of Stephen Hawking delivered a public lecture via his now-familiar voice synthesizer. ("Pardon my American accent," Hawking likes to quip.) The eleven-hundred-seat auditorium was easily filled, while an additional four hundred viewed a video feed in another hall, and hundreds more on the lawn squatted and listened to theater speakers broadcasting this scientific saint's epistle to the apostles.

The lecture was slated for 8:00 p.m. By three in the afternoon a line began to snake around the grassy quad adjoining the auditorium. By five o'clock hundreds of rank-and-file NASA JPL scientists flipped Frisbees and shared cooler drinks with students from Caltech and surrounding universities. Passersby might have been forgiven for assuming Bono or Britney was performing that night.

When he rolled into the auditorium and down the aisle in his motorized wheelchair, everyone arose in applause—a "standing O" just for showing up! The sermon was his customary scientific one on the Big Bang, black holes, time, and the universe, with the theology coming in the question-and-answer period. Although a few were interested in the minutiae of quantum mechanics, string theory, and inflationary cosmology, what most wanted were Deep Answers to Big Questions: "How did time begin?" "What was there before the Big Bang?" "Why does the universe exist?" Although Hawking has dared to attempt answers to such questions, some people sought even more. Here was an opportunity to inquire of a transcendent mind the biggest question of all: "Is there a God?"

Asked this ultimately unanswerable question, Hawking sat rigidly in his chair, stone quiet, his eyes darting back and forth across the computer screen. A minute, maybe two, went by as cosmologist Kip Thorne patiently explained how Stephen's computer—and his brain—work. Finally, after what seemed like an eternity, a wry smile formed and the Delphi oracle spoke: "I do not answer God questions."

What is it about Hawking that draws us to him as a scientific saint? He is, I believe, the embodiment of a larger social phenomenon known as *scientism*. Scientism is a scientific worldview that encompasses natural explanations for all phenomena, eschews supernatural and paranormal speculations, and embraces empiricism and reason as the twin pillars of a philosophy of life appropriate for an Age of Science.

Scientism's voice can best be heard through a new genre of science writing at a level appropriate for both lay readers and professionals, and can be seen in the works of such scientists as Carl Sagan, Stephen Jay Gould, Richard Dawkins, Edward O. Wilson, Jared Diamond, and many others who write or wrote for the masses and the ages. Scientism is a bridge spanning the abyss between what physicist C. P. Snow famously called the "two cultures" of science and the arts/humanities (where neither encampment could communicate with the other). Scientism has generated a new literati and intelligentsia passionately interested in the profound philosophical, ideological, and theological implications of scientific discoveries.

Although the origins of the scientism genre can be traced to the writings of Galileo and Thomas Huxley in centuries past, its modern incarnation began in the 1960s with the mathematician Jacob Bronowski's *The Ascent of Man*, took off in the 1980s with Carl Sagan's *Cosmos*, and hit pay dirt in the 1990s with Hawking's *A Brief History of Time*, which set new sales standards with a record two hundred weeks on the *Sunday Times* of London's hardcover best-seller list, and sold more than 10 million

Figure 4-1. The author with Stephen Hawking

copies in thirty-plus languages worldwide. Hawking's latest work, *The Universe in a Nutshell*, is already riding high on the *New York Times* bestseller list.

Hawking's towering fame is a result of a concatenation of variables that include the power of the scientism culture through which he writes; his creative insights into the ultimate nature of the cosmos in which he dares to answer ersatz-theological questions; and, most notably, his unmitigated heroism in the face of near-insurmountable physical obstacles that would have felled a lesser being.

But Hawking's individual success in particular, and the rise of scientism in general, reveal something deeper still. First, not all sciences share equally in scientism's power structure—cosmology and evolutionary theory ask the ultimate origin questions that have traditionally, and exclusively, been the providence of religion and theology. Scientism is courageously proffering naturalistic answers that supplant supernaturalistic ones, and in the process provides spiritual sustenance for those whose needs are not being met by these ancient cultural traditions. Second, we are, at base, a social hierarchical primate species. We show deference to our leaders, pay respect to our elders, and follow the dictates of our shamans; since this is the Age of Science, it is scientism's shamans who command our veneration. Third, because of language we are also storytelling, mythmaking primates, with scientism as the foundational stratum of our story and scientists as the premier mythmakers of our age.

So on the occasion of Stephen Hawking's sixtieth birthday we celebrate both this scientific shaman and the glorious culture of scientism in which we live.

Update: Stephen Hawking turned seventy-three in 2015 and is still going strong at the time of this writing, cranking out papers and books and giving lectures and talks for colleagues and the general public, and still drawing massive audiences at Caltech.

5

The Physicist and the Abalone Diver

The differences between the creators of two new theories of science reveal the social nature of the scientific process

Consider the following quotes, written by two different authors from recently self-published books purporting to revolutionize science:

> This book is the culmination of nearly twenty years of work that I have done to develop that new kind of science. I had never expected it would take anything like as long, but I have discovered vastly more than I ever thought possible, and in fact what I have done now touches almost every existing area of science, and quite a bit besides. I have come to view [my discovery] as one of the more important single discoveries in the whole history of theoretical science.

> The development of this work has been a completely solitary effort during the past thirty years. As you will realize as you read through this book, these ideas had to be developed by an outsider. They are such a complete reversal of contemporary thinking that it would have been very difficult for any one part of this integrated theoretical system to be developed within the rigid structure of institutional science.

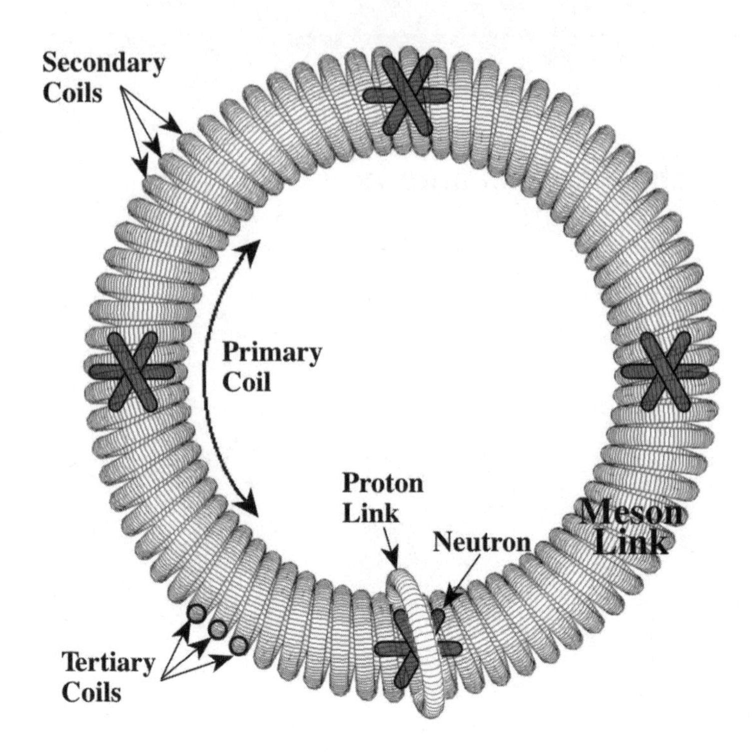

Figure 5-1. The Circlon Model of Nuclear Structure

James Carter's theory bases the structure of the entire universe—from atoms to galaxies—on the "circlon," "hollow, ring-shaped, mechanical particles that are held together within the nucleus by their physical shapes." From http://www.circlon.com

Both authors worked in isolation for decades. Both make equally extravagant claims about overturning the foundations of physics in particular and science in general. Both shunned the traditional route of peer-reviewed scientific papers and instead chose to take their ideas straight to the public through popular tomes. And both texts are filled with hundreds of self-produced diagrams and illustrations alleging to reveal the fundamental structures of nature.

There is one distinct difference between the two authors: one was featured in *Time*, *Newsweek*, *Wired*, and his book was reviewed in the *New York Times*. The other author has been completely ignored, with the exception of being featured in an exhibition in a tiny Southern California art museum. Their bios help clarify these rather different receptions.

One of the authors earned his PhD in physics at age twenty at Caltech,

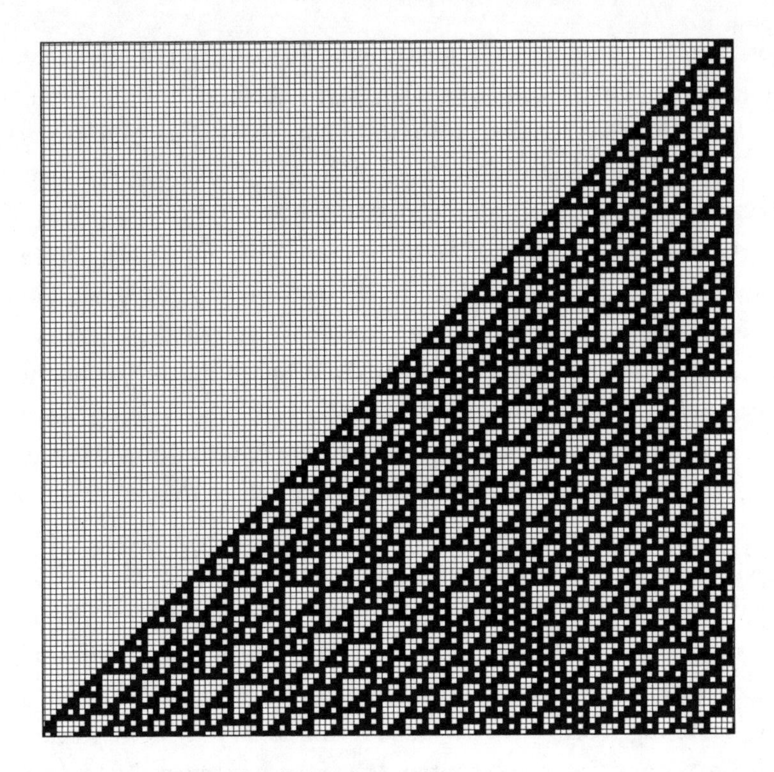

Figure 5-2. Stephen Wolfram's cellular automaton

Stephen Wolfram claims that the entire universe is a big cellular-automaton computer in which complexity can be generated from a handful of simple computational rules and algorithms. This cellular-automaton is from page 32 of Wolfram's book, *A New Kind of Science*, generated from rule number 110, which involves 150 steps that include how black and white squares interact with each other.

where Richard Feynman called him "astonishing," and was the youngest to ever win a prestigious MacArthur "genius" award. He founded an institute for the study of complexity at a major university, then quit to start his own software company, where he produced a wildly successful computer program used by millions of scientists and engineers. The other author is a former abalone diver, gold miner, filmmaker, cave digger, repairman, inventor, proprietor of a company that designs and builds underwater lift bags, and the owner-operator of a trailer park. Can you guess the authors and who penned which quote?

The first quote comes from Stephen Wolfram, the Caltech whiz and author of *A New Kind of Science*, in which the fundamental structure of the universe and everything in it is reduced to computational rules and algorithms that produce complexity in the form of cellular automata. The second quote comes from James Carter, the abalone diver and author of *The Other Theory of Physics*, proffering a "circlon" theory of the universe, where all matter is founded on these hollow, ring-shaped tubes that link everything together, from atoms to galaxies.

Whether Wolfram is right remains to be seen, but eventually we will find out because his ideas will be tested in the competitive marketplace of science. We will never know the veracity of Carter's ideas because they will never be taken seriously by scientists. Why? Because, like it or not, in science, as in most human intellectual endeavors, who is doing the saying matters as much as what is being said, at least in terms of getting an initial hearing. (If Wolfram is wrong his theory will go the way of the phlogiston, aether, and, well, the circlon.)

Science is, in this sense, conservative and sometimes elitist. It has to be in order to survive in a surfeit of would-be revolutionaries. Given limited time and resources, and the fact that for every Stephen Wolfram there are a hundred James Carters, one has to be selective. There needs to be some screening process whereby true revolutionary ideas are weeded out from ersatz ones. Enter the skeptics. We are interested in the James Carters of the world, in part because in studying how science goes wrong we learn how it can go right. But we also explore the interstices between science and pseudoscience because it is here where the next great revolution in science may arise. Although most of these ideas will join the phlogiston on the junk heap of science . . . you never know until you look more closely.

Update: James Carter's work on the circlon theory is online: www.circlon.com/ *The science writer Margaret Wertheim wrote a book about Carter and other "fringe scientists" titled* Physics on the Fringe: physicsonthefringe.com

6

A Candle in the Dark

Instead of cursing the darkness of pseudoscience on television, light a candle with Cable Science Network

Ever since Galileo began the tradition of communicating science in the vernacular so that all may share in its fruits, a tension has existed between those—call them *excluders*—who think science is for professionals only and regard its dissemination to wider audiences as *infra dig*, and those—call them *includers*—who understand that all levels of science require clear composition and public understanding of process and product.

Throughout much of the twentieth century the excluders have ruled the roost, punishing those in their flock who dared to write for those paying the bills. The Cornell University astronomer Carl Sagan, for example, whose PBS television series *Cosmos* was viewed by more than half a billion people, was denied membership in the National Academy of Science primarily (his biographers have demonstrated through interviews with insiders) because he invested too much time in science popularization.

Over the past two decades, however, a new literary genre has arisen in which professional scientists are presenting original research and theories in books written for both colleagues and the public. Most of Stephen Jay Gould's works are in this genre, as are those of Edward O. Wilson, Ernst Mayr, Jared Diamond, Richard Dawkins, Steven Pinker, and others. In fact, if you want to be considered a cultured person in today's society it is not enough to be steeped in literature, art, and music. You also need to know something about science.

The problem is that most people do not get their science through books and PBS documentary series. Although science junkies can fill their trough with such outstanding series as PBS's *Nova* and *Scientific American Frontiers*, most folks pick up bits and pieces of science from short newspaper articles or in evening news sound bites, which typically alternate between scary medical findings and stunning Hubble Space Telescope photographs, leaving out the subtleties of how science is really done and why contradictory findings do not mean that science has failed. Worse still, most networks pander to the ratings game and air a mélange of pseudoscience about ESP, UFOs, and moon landing hoaxes.

Like most scientists, I complain bitterly and often about such dismal programming. We write letters to network executives, but to no avail. One solution is to create our own network. A Cable Science Network, or CSN, in fact, is in the offing. Roger Bingham, of the Center for Brain and Cognition at the University of California, San Diego, is spearheading a movement (of which I am a part, along with Sagan's widow, Ann Druyan, and Salk Institute neuroscientist Terry Sejnowski) to launch a nonprofit organization modeled on the ubiquitous Cable-Satellite Public Affairs Network, or C-SPAN, now available in more than eighty million homes. CSN would be *science 24/7*—all science, all the time—freeing us, in Bingham's words, from "the tyranny of the sound bite."

Wouldn't it be great to watch congressional hearings on cloning, bioterrorism, global warming, and aging? Wouldn't it be fabulous to attend—via cable—cutting-edge lectures given by scientists at various annual scientific conferences? Every year tens of thousands of neuroscientists, for example, converge to exchange data on how the brain works. Wouldn't you love to sit in on some of those presentations instead of waiting to hear about one of them on television in a thirty-second encapsulation? I host a monthly science lecture series at Caltech, attended by more than three hundred people, who have seen lectures by such science luminaries as the aforementioned Pinker, Dawkins, and Gould. Wouldn't it be marvelous if three hundred thousand people could see those lectures?

With CSN all this, and more, will bring science to the people—and to scientists, legislators, teachers, and students—as never before. Sagan called science "a candle in the dark." CSN is still in the developmental stage, but if we can switch it on, it will be a candle whose light will illuminate a path toward the globalization of science.

Update: CSN is now TSN (http://thesciencenetwork.org/): *The Science Network, and has done remarkable work in making available online to millions a number of conferences, lectures, and interviews on a wide range of topics. Most memorable are the Beyond Belief conferences that have featured Richard Dawkins, Sam Harris, Neil deGrasse Tyson, Lawrence Krauss, Daniel Dennett, Sean Carroll, yours truly, and many others:* http://thescience network.org/programgroup/beyond-belief

7

The Feynman-Tufte Principle

A visual display of data should be simple enough to fit on the side of a van

I had long wanted to meet Edward R. Tufte—the man the *New York Times* called "the Leonardo da Vinci of data" because of his concisely written, artfully illustrated, and elegantly produced books on the visual display of data—and invite him to speak at the Skeptics Society science lecture series that I host at Caltech. But how could we afford someone of his stature? "My honorarium," he told me when I queried, "is to see Feynman's van."

The late Caltech physicist Richard Feynman is famous for many things—working on the atomic bomb at Los Alamos, winning a Nobel Prize in physics, cracking safes, playing drums, drawing nudes, and telling fantastic stories about his adventurous life (collected in *Surely You're Joking, Mr. Feynman*). Here in Pasadena, Feynman's fame extended to a 1975 Dodge Tradesman Maxivan outfitted with strange diagrams on the side panels. Most people who saw the van gazed in puzzlement at the squiggly lines, but once in a while someone would ask the driver why he had Feynman diagrams all over his van, only to be told "Because I'm Richard Feynman!"

Feynman first published his diagrams in a 1949 paper titled "Space-Time Approach to Quantum Electrodynamics," after which they became regular tools for physicists to use in computing the probabilities of quantum events. Feynman diagrams are simplified visual representations of the very complex world of quantum electrodynamics (QED), in which photons are depicted by wavy lines, electrons by straight or curved nonwavy lines, and at line junctions electrons emit or absorb a photon. Feynman thought in visual metaphors, such as this description of physical law: "Nature uses only the longest threads to weave her pattern, so each small piece of the fabric reveals the organization of the entire tapestry." As the physicist and Feynman colleague Freeman Dyson recalled, "When Feynman's tools first became available, it was a tremendous liberation. You could do all kinds of things with them you couldn't have done before." The physicist Hans Bethe concurred: "The great power of Feynman's diagrams is that they combine many steps of the older calculations in one."

In the diagram on the back door of the van, photographed here, time flows from bottom to top. If you cover it with a sheet of paper and pull up you will see a pair of electrons (the straight lines) moving toward each other. When the left-hand electron emits a photon (wavy line junction) it is deflected outward left; the photon is reabsorbed by the right-hand electron, causing it to deflect outward right. (The "Tuva or Bust" bumper sticker refers to Feynman's attempt to travel to the tiny Asian country of Tannu Tuva "because any place that's got a capital named K-Y-Z-Y-L has just got to be interesting!" He died before making it.) The second photograph shows Tufte photographing the van's additional Feynman diagrams, most showing more than one photon emission or absorption.

Feynman diagrams are the embodiment of what Edward Tufte teaches and preaches about analytical design in his seminars (www.edwardtufte .com) and books (*The Visual Display of Quantitative Information, Visual Explanations,* and *Envisioning Information*): "Good displays of data help to reveal knowledge relevant to understanding mechanism, process and dynamics, cause and effect." Humans are very visual primates, seeing the unthinkable and thinking the unseeable. Eye and brain are inseparable. "Visual representations of evidence should be governed by principles of reasoning about quantitative evidence. Clear and precise seeing becomes as one with clear and precise thinking."

The master of clear and precise thinking meets the master of clear and precise seeing in what I call the Feynman-Tufte Principle: *A visual display of data should be simple enough to fit on the side of a van.*

Figure 7-1.

The author and Edward Tufte with Feynman's van

As Tufte poignantly demonstrated in his analysis of the space shuttle *Challenger* disaster, "a clear proximate cause [was] an inability to assess the link between cool temperature and O-ring damage on earlier flights" in the thirteen charts prepared by Thiokol (makers of the solid rocket booster that blew up) for NASA. The loss of the *Columbia*, Tufte believes, was directly related to "a PowerPoint festival of bureaucratic hyperrationalism" in which a single slide contained six different levels of hierarchy, thereby obfuscating the conclusion that damage to the left wing might be significant. As Tufte notes, in his 1963 classic work *The Feynman Lectures on Physics*, Feynman covered all of physics—from celestial mechanics to quantum electrodynamics—with only two levels of hierarchy: chapters and chapter subheads.

Tufte codified the design process into six principles: "(1) *documenting* the sources and characteristics of the data, (2) insistently enforcing appropriate *comparisons*, (3) demonstrating mechanisms of *cause and effect*, (4) expressing those mechanisms *quantitatively*, (5) recognizing the inherently *multivariate* nature of analytic problems, (6) inspecting and evaluating *alternative explanations*." In brief, "information displays should be documentary, comparative, causal and explanatory, quantified, multivariate, exploratory, skeptical."

Skeptical. How fitting for this column, because when I asked Tufte to summarize the goal of his work, he said, "simple design, intense content." Since we all need a mark at which to aim (one meaning of "skeptic"), *simple design, intense content* is a sound objective for the future of this series.

Update: Tufte went on to produce an art exhibition at Fermilab—a particle accelerator in Chicago—featuring Feynman diagrams and even Feynman's van, which has now been restored by Seamus Blackley and Ralph Leighton. http://bit.ly/1nuSlEn

8

The Flipping Point

How the evidence for anthropogenic global warming has converged to cause this environmental skeptic to make a cognitive flip

In 2001 Cambridge University Press published Bjorn Lomborg's book *The Skeptical Environmentalist*, which I thought was a perfect debate topic at the Skeptics Society public lecture series at the California Institute of Technology. The problem was that all of the top environmental organizations refused to participate. "There is no debate," one told me. "We don't want to dignify that book," said another. One leading environmentalist warned me that my reputation would be irreparably harmed if I went through with it. So of course I did.

My experience is symptomatic of deep problems that have long plagued the environmental movement. Activists who vandalize Hummer dealerships and destroy logging equipment are criminal ecoterrorists. Environmental groups who cry doom and gloom to keep donations flowing only hurt their credibility. As an undergraduate in the 1970s I learned (and believed) that by the 1990s overpopulation would lead to worldwide starvation and the exhaustion of key minerals, metals, and oil, predictions that failed utterly. Politics polluted the science and made me an environmental skeptic.

Nevertheless, data trump politics, and a convergence of evidence from numerous sources has led me to make a cognitive switch on the subject of anthropogenic global warming. My attention was piqued on February 8, 2006, when eighty-six leading evangelical Christians—the last cohort I expected to get on the environmental bandwagon—issued the Evangelical Climate Initiative calling for "national legislation requiring economy-wide reductions" in carbon emissions. After attending a 2002 Oxford conference on the science of global warming, the chief lobbyist for the National Association of Evangelicals, the Reverend Richard Cizik, described his experience as "a conversion . . . not unlike my conversion to Christ."

Then I attended the TED (Technology, Entertainment, Design) conference in Monterey, California, where former vice president Al Gore delivered the single finest summation of the evidence for global warming I have ever heard, based on the 2006 documentary film about his work in this area, *An Inconvenient Truth.* Because we are primates with such visually dominant sensory systems we need to *see* the evidence to believe it, and the striking before-and-after photographs showing the disappearance of glaciers around the world shocked me out of my skepticism.

Four books then took me to the flipping point. Archaeologist Brian Fagan's *The Long Summer* (Basic Books, 2004) documents how civilization is the gift of a temporary period of mild climate. Geographer Jared Diamond's *Collapse* (Viking, 2005) demonstrates how natural and human-caused environmental catastrophes led to the collapse of civilizations. Journalist Elizabeth Kolbert's *Field Notes from a Catastrophe* (Simon & Schuster, 2006) is a page-turning account of her journeys around the world with environmental scientists who are documenting species extinction and climate change that are unmistakably linked to human action. And biologist Tim Flannery's *The Weather Makers* (Atlantic Monthly Press, 2006) reveals how he went from being a skeptical environmentalist to a believing activist as incontrovertible data linking the increase of carbon dioxide, CO_2, to global warming accumulated in the previous decade.

It is a matter of CO_2 Goldilocks. In the last ice age, CO_2 levels were 180 parts per million (ppm)—too cold. Between the Agricultural Revolution and the Industrial Revolution, CO_2 levels rose to 280 ppm—just right. Today CO_2 levels are at 380 ppm and are projected to reach 450 to 550 ppm by the end of the century—too warm. Like a kettle of water that transforms from liquid to steam when it changes from 211 to 212 degrees F, the environment itself is about to make a CO_2-driven flip.

According to Flannery, even if we reduce our CO_2 emissions by 70 percent by 2050, average global temperatures will increase between 2 to 9 degrees C by 2100. This rise could lead to the melting of the Greenland Ice Sheet, which the March 24 issue of *Science* reports is already shrinking at a rate of 224±41 cubic kilometers per year, double the rate measured in 1996 (Los Angeles uses 1 cubic kilometer of water per year). If it and the West Antarctic Ice Sheet melt, sea levels will rise 5 to 10 meters, displacing half a billion inhabitants of coastal communities.

Because of the complexity of the problem, environmental skepticism was once tenable. No longer. It is time to flip from skepticism to activism.

Update: I wrote another column, in the August 2014 edition of Scientific American (http://bit.ly/1sS7K3y), *in which I maintained my belief that climate change is real and human caused, but made the argument that there are other problems in the world equally pressing but more immediate, such as poverty, disease, and starvation. Not surprisingly, I received almost as much hate mail for that column as I did for this one. With the possible exception of religion and politics (and I'm not even sure about that), there is no more contentious topic for a science writer to touch today than climate change, regardless of where you come down on it.*

9

Fake, Mistake, Replicate

A court of law may determine the meaning of replication in science

In the rough-and-tumble world of science, disputes are usually settled in time, as a convergence of evidence accumulates in favor of one hypothesis over another. Until now.

On April 10, 2006, the economist John Lott filed a defamation lawsuit against the economist Steven Levitt and HarperCollins, the publisher of Levitt's 2005 book *Freakonomics.* At issue is what Levitt meant when he wrote that scholars could not "replicate his results," referring to Lott's 1998 book *More Guns, Less Crime* (University of Chicago Press). Lott employed a sophisticated statistical analysis on data from state-level variation in "carry and conceal" laws, finding that states that passed laws that permit citizens to carry concealed weapons saw statistically significant declines in robbery, rape, and homicide compared with states that did not pass such laws.

As is typical with such politically charged research, considerable controversy followed publication of Lott's book, with a flurry of conference presentations and journal papers published, some of which replicated his results and some of which did not. For example, in a series of papers published in the Yale and Stanford law reviews (available at http://papers.ssrn.com), Lott and his critics debated the evidence.

In *Freakonomics*, Levitt proffered his own theory for the 1990s crime decline—*Roe v. Wade*. According to Levitt, children born into impoverished and adverse environments are more likely to grow up to become criminals. After *Roe v. Wade*, millions of poor, single women had abortions instead of future potential criminals; twenty years later the pool of potential criminals had shrunk, along with the crime rate. Levitt employed a comparative statistical analysis to show that the five states that legalized abortion two years before *Roe v. Wade* witnessed a crime fall earlier than the other forty-five states. Further, those states with the highest abortion rates in the 1970s experienced the greatest fall in crime in the 1990s. Finally, Levitt showed that three additional factors reduced the pool of criminals: increased rates of imprisonment, increased number of police, and the bursting of the crack cocaine bubble.

One factor that Levitt dismissed is Lott's, in a single passage in the middle of a thirty-page chapter, in which he concluded that "Lott's admittedly intriguing hypothesis doesn't seem to be true. When other scholars have tried to replicate his results, they found that right-to-carry laws simply don't bring down crime."

According to Lott's legal complaint, "the term 'replicate' has an objective and factual meaning" and that other scholars "have analyzed the identical data that Lott analyzed and analyzed it the way Lott did in order to determine whether they can reach the same result." When Levitt said that they could not, he is "alleging that Lott falsified his results."

I asked Levitt what he meant by replicate. "I used the term in the same way that most scientists do—substantiate results." Substantiate, not duplicate. Did he mean to imply that Lott falsified his results? "No I did not." In fact, others have accused Lott of falsifying his data, so I asked him why he is suing Levitt. "Having some virtually unheard of people making allegations on the Internet is one thing. Having claims made in a book published by an economics professor and printed by a reputable book publisher, already with sales exceeding a million copies, is something entirely different. In addition, Levitt is well-known and his claims unfortunately carry some weight. I have had numerous people ask me after reading *Freakonomics* whether it is really true that others have been unable to replicate my research."

"Replicate" is a verb that depends on the sentence's subject. "Replicate methodology" might capture Lott's meaning, but "replicate results" means testing the conclusion of the methodology, in this case that more guns causes less crime. The problem is that many scientific experiments and statistical data sets are so complicated that the failure to replicate more likely indicates unconscious mistakes made during the original research or in the replication process, rather than conscious fakery.

Mr. Lott, tear down this legal wall and let us return to doing science without lawyers. Replicating results means testing hypotheses, not merely duplicating methodologies, and this central tenet of science can only flourish in an atmosphere of open peer review.

Update: A federal judge found in Levitt's favor in the charge that his claim that scholars had not replicated Lott's findings in Freakonomics *was not defamatory; however, the judge found in Lott's favor in a portion of the complaint involving a private email that Levitt sent to another economist about Lott's research, leading to Levitt issuing a retraction letter. See* http://bit.ly/1lC5AH5 *and* http://bit.ly/Y2j70a

10

Wronger Than Wrong

Not all wrong theories are equal

In belles lettres the witty literary slight has evolved into a genre because, as the twentieth-century trial lawyer Louis Nizer noted, "A graceful taunt is worth a thousand insults." To wit from high culture, Samuel Johnson: "He is not only dull himself, he is the cause of dullness in others." Mark Twain: "I didn't attend the funeral, but I sent a nice letter saying I approved of it." Winston Churchill: "He has all the virtues I dislike and none of the vices I admire." And from pop culture, Groucho Marx: "I've had a perfectly wonderful evening. But this wasn't it."

Scientists are no slouches when it comes to pitching clever invectives at colleagues. Achieving almost canonical status as the ne plus ultra scientific put-down is the theoretical physicist Wolfgang Pauli's withering critique of a paper: "This isn't right. It's not even wrong." I call this Pauli's Proverb.

The Columbia University mathematician Peter Woit recently employed Pauli's Proverb in his book title, a critique of string theory called *Not Even Wrong* (Basic Books, 2006). String theory, Woit argues, is not only based on nontestable hypotheses, it also depends far too much on the aesthetic nature of its mathematics and the eminence of its proponents. In science, if an idea is not falsifiable, it is not that it is wrong; it is that we cannot determine if it is wrong, and thus it is not even wrong.

Not even wrong. What could be worse? Being wronger than wrong, or what I call Asimov's Axiom, well stated in his book *The Relativity of Wrong* (Doubleday, 1988):

> When people thought the earth was flat, they were wrong. When people thought the earth was spherical, they were wrong. But if you think that thinking the earth is spherical is just as wrong as thinking the earth is flat, then your view is wronger than both of them put together.

Aximov's Axiom holds that science is cumulative and progressive, building on the mistakes of the past, and that even though scientists are often wrong, their wrongness attenuates with continued data collection and theory building.

The view that all wrong theories are equal implies that no theory is better than any other. This itself is a theory known as the "strong social construction of science," which contends that science is inextricably bound to the social, political, economic, religious, and ideological predilections of a culture, particularly of those in power. Scientists are knowledge capitalists who produce scientific papers that report the results of experiments conducted to test (and usually support) the hegemonic theories that reinforce the status quo.

In some extreme cases, particularly in the social sciences, this theory is right. In the early nineteenth century, physicians discovered that slaves suffered from *drapetomania*, or the uncontrollable urge to escape from slavery, and *dysathesia aethiopica*, or the tendency to be disobedient to slave masters. In the late nineteenth and early twentieth centuries, scientific measurements of racial differences in cognitive abilities found that blacks were inferior to whites. In the mid-twentieth century psychiatrists discovered evidence that led them to classify homosexuality as a disease. And until recently, women were considered inherently inferior in science classrooms and corporate boardrooms.

Such egregious examples, however, do not negate the extraordinary ability of science to elucidate the natural and social worlds. Reality exists, and science is the greatest tool ever employed to discover and describe that reality. The theory of evolution, even though experiencing vigorous debates about the tempo and mode of life's history, is vastly superior to the theory of creation, which is not even wrong (in Pauli's Proverb sense). As Richard Dawkins observed on this dispute: "When two opposite points of view are expressed with equal intensity, the truth does not necessarily lie exactly halfway between them. It is possible for one side to be simply wrong."

Simply wrong. When people thought that science was unbiased and unbound by culture, they were simply wrong. When people thought that science was completely socially constructed, they were simply wrong. But if you think that thinking science is unbiased is just as wrong as thinking that science is socially constructed, then your view is not even wronger than wrong.

II

SKEPTICISM

11

Fox's Flapdoodle

Tabloid television offers a lesson in uncritical thinking

The price to pay for liberty, in addition to eternal vigilance, is eternal patience with the vacuous blather occasionally expressed beneath the shield of free speech. It is a cost worth bearing, but it does become exasperating now and again, as on February 15, 2001, when Fox aired its highly advertised special "Conspiracy Theory: Did We Land on the Moon?" NASA, we are told, faked the whole thing on a movie set.

Such flummery should not warrant wasting precious time in responding, but in a free society skeptics are the watchdogs of irrationalism—the consumer advocates of bad ideas. Yet debunking is not simply the divestment of bunk; its utility is found in offering a viable alternative, along with a lesson on how thinking goes wrong. The Fox show is a case study, starting with their disclaimer: "The following program deals with a controversial subject. The theories expressed are not the only possible explanation. Viewers are invited to make a judgment based on all available information." That information, of course, was not provided, so let's go through this point by point just in case the statistic at the top of the show—20 percent of Americans believe we never went to the moon—is accurate.

Claim: There are two sources of light in the photographs taken on the moon. Since there is only one source of light in the sky (the sun) the extra "fill" light comes from studio spotlights.

Answer: Setting aside the inane assumption that NASA and their coconspirators were too incogitant to have thought of this, there are actually *three* sources of light: the sun, the earth reflecting the sun, and the moon itself, which acts as a powerful reflector, particularly when you are standing on it.

Claim: The American flag was "waving" in the airless environment of the moon, as clearly seen in the footage as the astronaut plants it.

Answer: The flag was "waving" while the astronaut was fiddling with it back and forth as he jammed it into the hole, but the moment he let go of it the waving stopped. The flag remained stiff because NASA sewed a metal rod through the top of it.

Claim: There was no blast crater beneath the LEM lander.

Answer: There are only a couple of inches of moon dust, beneath which is a solid surface that would not be affected by the blast of the LEM engine.

Claim: When the top half of the LEM took off from the moon there was no rocket flame like we see on earth. The LEM just seems to leap off the base like it was yanked up by cables.

Answer: First, you can clearly see in the footage of the launch that there *is* quite a blast as dust and other particles go flying. Second, since there is no oxygen there is no fuel to generate a rocket nozzle flame tail.

Claim: The LEM simulator used by astronauts for practice was obviously unstable—Neil Armstrong barely escaped with his life as his simulator crashed and he ejected just seconds before impact. The real LEM was much larger and heavier and thus impossible to land.

Answer: Practice makes perfect, and these guys practiced and practiced. Analogously, a bicycle is inherently unstable until you learn to ride it. Plus the moon's gravity is only a sixth that of the earth, so weight is less of a factor.

Claim: There are no stars in the sky in the photographs and films from the moon.

Answer: There are no stars in the sky in photographs and films shot on the earth either. If you want to photograph stars in the night sky you have to leave the shutter open for a while. Stars are simply too faint to appear on film when shot at normal speed.

Claim: The Van Allen radiation belts surrounding the earth would have fried the astronauts with a lethal dose of radiation.

Answer: If you pass through the Van Allen belts reasonably fast there is not enough radiation to cause harm, which is what the Apollo spacecraft did.

The no-moonie conspiracy mongers go on and on in this vein, weaving their narratives that include the "murder" of numerous astronauts and pilots in "accidents," including Gus Grissom in the *Apollo I* fire before he was about to go public with the hoax. The lesson in all this fiddle-faddle is that, like most conspiracy theories, there is no positive evidence in support, only negative evidence in the form of "they covered it up." I once asked G. Gordon Liddy (who should know) about conspiracies. He quoted *Poor Richard's Almanack*: "Three people can keep a secret if two of them are dead." To think that thousands of NASA scientists would keep their mouths shut for years is risible rubbish.

So, two thumbs down to Fox for their blatant quest for ratings through tabloid trash television, and two thumbs up to NASA for solving seemingly insoluble problems to get humans to the moon. Now we should be planting flags and taking photographs on Mars . . . and beyond. *Ad astra!*

Update: As noted in essay number 3 in this volume, the additional reason there is no rocket flame coming out of the LEM engine is that it used hypergolic propellants (dinitrogen tetroxide and Aerozine 50) that ignite upon contact and burn very cleanly. The most entertaining response to the "no moonies" I've seen is that of Buzz Aldrin, who punched out Bart Sibrel after being harassed by him, captured on video: http://bit.ly/1nrhrER

12

Baloney Detection

How to draw boundaries between science and pseudoscience, Part I

When lecturing on science and pseudoscience at colleges and universities, I am inevitably asked, after challenging common beliefs held by many students, "Why should we believe you?" My answer: "You shouldn't."

I then explain that we need to check things out for ourselves and, short of that, at least to ask basic questions that get to the heart of the validity of any claim. This is what I call baloney detection, in deference to Carl Sagan, who coined the phrase "Baloney Detection Kit." To detect baloney—that is, to help discriminate between science and pseudoscience—I suggest ten questions to ask when encountering any claim.

1. How reliable is the source of the claim? As Daniel Kevles showed so effectively in his book *The Baltimore Affair*, in investigating possible scientific fraud there is a boundary problem in detecting a fraudulent signal within the background noise of mistakes and sloppiness that is a normal part of the scientific process. The investigation of research notes in a laboratory affiliated with Nobel laureate David Baltimore by an independent committee established by Congress to investigate potential fraud revealed a surprising number of mistakes. But science is messier than most people realize. Baltimore was exonerated when it became clear that there was no purposeful data manipulation.

2. Does this source often make similar claims? Pseudoscientists have a habit of going well beyond the facts, so when individuals make numerous extraordinary claims they may be more than just iconoclasts. This is a matter of quantitative scaling, since some great thinkers often go beyond the data in their creative speculations. Cornell's Thomas Gold is notorious for his radical ideas, but he has been right often enough that other scientists listen to what he has to say. Gold proposes, for example, that oil is not a fossil fuel at all, but the by-product of a deep, hot biosphere. Hardly any earth scientists I have spoken with take this thesis seriously, yet they do not consider Gold a crank. What we are looking for here is a pattern of fringe thinking that consistently ignores or distorts data.

3. Have the claims been verified by another source? Typically pseudoscientists will make statements that are unverified, or verified by a source within their own belief circle. We must ask who is checking the claims, and even who is checking the checker. The biggest problem with the cold fusion debacle, for example, was not that Stanley Pons and Martin Fleischman were wrong; it was that they announced their spectacular discovery before it was verified by other laboratories (at a press conference, no less), and, worse, when cold fusion was not replicated, they continued to cling to their claim.

4. How does the claim fit with what we know about how the world works? An extraordinary claim must be placed into a larger context to see how it fits. When people claim that the pyramids and the sphinx were built more than ten thousand years ago by an advanced race of humans, they are not presenting any context for that earlier civilization. Where are the rest of the artifacts of those people? Where are their works of art, their weapons, their clothing, their tools, their trash? This is simply not how archaeology works.

5. Has anyone gone out of the way to disprove the claim, or has only confirmatory evidence been sought? This is the *confirmation bias*, or the tendency to seek confirming evidence and reject or ignore disconfirming evidence. The confirmation bias is powerful and pervasive and is almost impossible for any of us to avoid. It is why the methods of science that emphasize checking and rechecking, verification and replication, and especially attempts to falsify a claim, are so critical.

In chapter 13 I will expand the baloney detection process with five more questions that reveal how science works to detect its own baloney.

13

More Baloney Detection

How to draw boundaries between science and pseudoscience, Part II

When exploring the borderlands of science, we often face a "boundary problem" of where to draw the line between science and pseudoscience. The boundary is the line of demarcation between geographies of knowledge, the border defining countries of claims. Knowledge sets are fuzzier entities than countries, however, and their edges are blurry. It is not always clear where to draw the line.

In chapter 12 I suggested five questions to ask about a claim to determine whether it is legitimate or baloney. Continuing with the baloney-detection questions, we see that in the process we are also helping to solve the boundary problem of where to place a claim.

6. Does the preponderance of evidence converge to the claimant's conclusion, or a different one? The theory of evolution, for example, is proven through a convergence of evidence from a number of independent lines of inquiry. No one fossil, no one piece of biological or paleontological evidence has "evolution" written on it; instead there is a convergence of evidence from tens of thousands of evidentiary bits that adds up to a story of the evolution of life. Creationists conveniently ignore this convergence, focusing instead on trivial anomalies or currently unexplained phenomena in the history of life.

7. Is the claimant employing the accepted rules of reason and tools of research, or have these been abandoned in favor of others that lead to the desired conclusion? UFOlogists suffer this fallacy in their continued focus on a handful of unexplained atmospheric anomalies and visual misperceptions by eyewitnesses while conveniently ignoring the fact that the vast majority (90 to 95 percent) of UFO citings are fully explicable with prosaic answers.

8. Has the claimant provided a different explanation for the observed phenomenon, or is it strictly a process of denying the existing explanation? This is a classic debate strategy—criticize your opponent and never affirm what you believe in order to avoid criticism. But this stratagem is unacceptable in science. Big Bang skeptics, for example, ignore the convergence of evidence of this cosmological model, focus on the few flaws in the accepted model, and have yet to offer a viable cosmological alternative that carries a preponderance of evidence in favor of it.

9. If the claimant has proffered a new explanation, does it account for as many phenomena as the old explanation? The HIV-AIDS skeptics argue that lifestyle, not HIV, causes AIDS. Yet, to make this argument they must ignore the convergence of evidence in support of HIV as the causal vector in AIDS, and simultaneously ignore such blatant evidence as the significant correlation between the rise in AIDS among hemophiliacs shortly after HIV was inadvertently introduced into the blood supply. On top of this, their alternative theory does not explain nearly as much of the data as the HIV theory.

10. Do the claimants' personal beliefs and biases drive the conclusions, or vice versa? All scientists hold social, political, and ideological beliefs that could potentially slant their interpretations of the data, but how do those biases and beliefs affect their research? At some point, usually during the peer-review system, such biases and beliefs are rooted out, or the paper or book is rejected for publication. This is why one should not work in an intellectual vacuum. If you don't catch the biases in your research, someone else will.

Clearly, there are no foolproof methods of detecting baloney or drawing the boundary between science and pseudoscience. Yet there is a solution: science deals in fuzzy fractions of certainties and uncertainties, where evolution and Big Bang cosmology may be assigned a 0.9 probability of being true, and creationism and UFOs a 0.1 probability of being true. In between are borderland claims: we might assign superstring theory a 0.7 and cryonics a 0.2. In all cases, we remain open-minded and flexible, willing to reconsider our assessments as new evidence arises. This is, undeniably, what makes science so fleeting and frustrating to many people; it is, at the same time, what makes science the most glorious product of the human mind.

14

Hermits and Cranks

Fifty years ago Martin Gardner launched the modern skeptical movement. Unfortunately, much of what he wrote about is still current today

In 1950 Martin Gardner published an article in the *Antioch Review* titled "The Hermit Scientist," about what we would today call pseudoscientists. It was Gardner's first publication of a skeptical nature (he was the games columnist for *Scientific American* for more than a quarter century), and in 1952 he expanded it into a book titled *In the Name of Science*, with the descriptive subtitle "An Entertaining Survey of the High Priests and Cultists of Science, Past and Present." Published by Putnam, the book sold so poorly that it was quickly remaindered and lay dormant until 1957, when it was republished by Dover and has come down to us as *Fads and Fallacies in the Name of Science*, still in print and arguably *the* skeptic classic of the past half century.

The "hermit scientist," a youthful Gardner wrote half a century ago, works alone and is ignored by mainstream scientists. "Such neglect, of course, only strengthens the convictions of the self-declared genius." Gardner, however, was wrong by half in his prognostications: "The current flurry of discussion about Velikovsky and Hubbard will soon subside, and their books will begin to gather dust on library shelves." While Velikovskians are a quaint few surviving in the interstices of fringe

Figure 14-1.
Martin Gardner's *In the Name of Science*, the "bible" of the modern skeptical movement.

culture, L. Ron Hubbard has been canonized by the Church of Scientology and deified as the founding saint of a world religion.

In 1952 Gardner could not have known that the nascent flying saucer craze would turn into an alien industry: "Since flying saucers were first reported in 1947, countless individuals have been convinced that the earth is under observation by visitors from another planet." Absence of evidence then was no more a barrier to belief than it is today, and UFOlogists proffered the same conspiratorial explanations for the dearth of proof: "I have heard many readers of the saucer books upbraid the government in no uncertain terms for its stubborn refusal to release the 'truth' about the elusive platters. The administration's 'hush-hush policy' is angrily cited as proof that our military and political leaders have lost all faith in the wisdom of the American people."

From his perspective half a century ago Gardner was even then bemoaning the fact that some beliefs never seem to go out of vogue, as he recalled H. L. Mencken's quip from the 1920s that "if you heave an egg out of a Pullman car window anywhere in the United States you are likely to hit a fundamentalist." Gardner cautions that when presumably religious superstition should be on the wane how easy it is "to forget that thousands of high school teachers of biology, in many of our southern states, are still afraid to teach the theory of evolution for fear of losing their jobs." Today, bleeding Kansas enjoins the fight as the creationist virus spreads northward.

Thankfully there has been some progress since Gardner published his first criticisms of pseudoscience. Now largely antiquated are Gardner's chapters on believers in a flat earth, a hollow earth, Velikovsky, Atlantis and Lemuria, Alfred William Lawson, Roger Babson, Trofim Lysenko, Wilhelm Reich, and Alfred Korzybski. But, disturbingly, a good two-thirds of the book's contents are relevant today, including homeopathy, naturopathy, osteopathy, iridiagnosis (reading the iris of the eye to determine bodily malfunctions), food faddists, cancer cures and other forms of medical quackery, Edgar Cayce, the great pyramid's alleged mystical powers, handwriting analysis, ESP and PK, reincarnation, dowsing rods, eccentric sexual theories, and theories of group racial differences.

And the motives of the "hermit scientists" have not changed either. Gardner recounts the day that Groucho Marx interviewed Louisiana state senator Dudley J. LeBlanc about a "miracle" cure-all vitamin and mineral tonic called Hadacol that the senator had invented. When Groucho asked him what it was good for, LeBlanc answered with uncharacteristic honesty: "It was good for five and a half million for me last year."

What I find especially valuable about Gardner's views are his insights into the differences between science and pseudoscience. On the one extreme we have ideas that are most certainly false, "such as the dianetic view that a one-day-old embryo can make sound recordings of its mother's conversation." In the borderlands middle "are theories advanced as working hypotheses, but highly debatable because of the lack of sufficient data," and for this Gardner selects a most propitious example: "the theory that the universe is expanding." That theory would now fall onto the spectrum at the other extreme end of "theories almost certainly true, such as the belief that the earth is round or that men and beasts are distant cousins."

How can we tell if someone is a scientific crank? Gardner offers this advice: (1) "First and most important of these traits is that cranks work in almost total isolation from their colleagues." Cranks typically do not understand how the scientific process works—that they need to try out their ideas on colleagues, attend conferences, and publish their hypotheses in peer-reviewed journals before announcing to the world their startling discovery. Of course, when you explain this to them they say that their ideas are too radical for the conservative scientific establishment to accept. (2) "A second characteristic of the pseudo-scientist, which greatly strengthens his isolation, is a tendency toward paranoia," which manifests itself in several ways:

> (1) He considers himself a genius. (2) He regards his colleagues, without exception, as ignorant blockheads. (3) He believes himself unjustly persecuted and discriminated against. The recognized societies refuse to let him lecture. The journals reject his papers and either ignore his books or assign them to "enemies" for review. It is all part of a dastardly plot. It never occurs to the crank that this opposition may be due to error in his work. (4) He has strong compulsions to focus his attacks on the greatest scientists and the best-established theories. When Newton was the outstanding name in physics, eccentric works in that science were violently anti-Newton. Today, with Einstein the father symbol of authority, a crank theory of physics is likely to attack Einstein. (5) He often has a tendency to write in a complex jargon, in many cases making use of terms and phrases he himself has coined.

We should keep these criteria at the forefront when we explore controversial ideas on the borderlands of science. "If the present trend continues," Gardner concludes, "we can expect a wide variety of these men, with theories yet unimaginable, to put in their appearance in the years immediately ahead. They will write impressive books, give inspiring lectures, organize exciting cults. They may achieve a following of one—or one million. In any case, it will be well for ourselves and for society if we are on our guard against them." So we still are, Martin. That is what skeptics do, and in tribute for all you have done we shall continue to honor your founding command.

15

Skepticism as a Virtue

An inquiry into the original meaning of the word "skeptic"

Poets often express deep insights into human nature in far less verbiage than scientists. Alexander Pope's *Essay on Man*, for example, is filled with pithy observations containing profound implications. Consider his description of the dualistic tensions in the human condition:

> *Place on this isthmus of a middle state,*
> *A being darkly wise and rudely great . . .*
> *He hangs between; in doubt to act or rest;*
> *In doubt to deem himself a god, or beast;*
> *In doubt his mind or body to prefer;*
> *Born but to die, and reasoning but to err.*

Pope has packed a lot into this refrain—humans as wise and rude, dark and great, god and beast—but the final clause is an important challenge to science: is all our reasoning for naught, to end only in error? Such fear haunts us in our quest for understanding, and it is precisely why skepticism is a virtue. We must always be on guard against errors in our reasoning. Eternal vigilance is the watchphrase not just of freedom, but also of thinking. That is the very nature of skepticism.

To my considerable chagrin it was five years into the editing and publishing of *Skeptic* magazine that I realized I had never bothered to define the word, or even examined how others had used it, until Stephen Jay Gould, in the foreword to my book *Why People Believe Weird Things*, mentioned that it comes from the Greek *skeptikos*, for "thoughtful." Etymologically, in fact, its Latin derivative is *scepticus*, for "inquiring" or "reflective." Further variations in the ancient Greek include "watchman" and "mark to aim at." Hence, skepticism is thoughtful and reflective inquiry. To be skeptical is to aim toward a goal of critical thinking. Skeptics are the watchmen of reasoning errors, the Ralph Naders of bad ideas.

This is a far cry from modern misconceptions of the word as "cynical" or "nihilistic," although a history of the word gives us some insight into how "skepticism" has become synonymous with "cynicism" or "nihilism." The *Oxford English Dictionary*, our finest source for historical word usage, offers as its first definition of "skeptic": "One who, like Pyrrho and his followers in Greek antiquity, doubts the possibility of real knowledge of any kind; one who holds that there are no adequate grounds for certainty as to the truth of any proposition whatever." This may be true in philosophy, but it certainly is not true in science. There are more than adequate grounds for the probability of the truth of propositions, but here I play with words a bit, substituting "probability" for "certainty," because there are no uncontrovertible facts in science, if we define fact as a belief held with 100 percent certitude. I cannot improve on Gould's definition of a fact in science: "confirmed to such a degree that it would be perverse to withhold provisional assent."

Superstring theory may be uncertain, but heliocentrism is not. Whether the history of life is best described by gradualism or punctuated equilibrium may still be in dispute, but the fact that life has evolved is not. The difference is one of probabilities, and this is reflected in a second usage of the word "skeptic": "One who doubts the validity of what claims to be knowledge in some particular department of inquiry." Okay, so we don't doubt everything, just some things, particularly those lacking in evidence and logic. Unfortunately, it is also true that some skeptics fall into a third usage of the word: "one who is habitually inclined rather to doubt than to believe any assertion or apparent fact that comes before him; a person of sceptical temper." Why some people are, by temperament, more skeptical than others is a subject for another essay, but suffice it to say that the reverse is also true—some folks are, by temperament, habitually inclined to believe rather than to doubt any assertion. Neither extreme is healthy and both lead to reasoning errors.

Perhaps the closest fit of the word "skeptic" to what we equate with a skeptical or scientific attitude is a fourth meaning: "a seeker after truth; an inquirer who has not yet arrived at definite convictions." Skepticism is not "seek and ye shall find"—a classic case of what is called the *confirmation bias* in cognitive psychology—but "seek and keep an open mind." What does it mean to have an open mind? It is to find the essential balance between orthodoxy and heresy, between a total commitment to the status quo and the blind pursuit of new ideas, between being open-minded enough to accept radical new ideas and so open-minded that your brains fall out. Skepticism is about finding that balance.

16

The Exquisite Balance

Science helps us understand the essential tension between orthodoxy and heresy in science

In a 1987 lecture, "The Burden of Skepticism," astronomer Carl Sagan succinctly summarized the essential tension between orthodoxy and heresy:

> It seems to me what is called for is an exquisite balance between two conflicting needs: the most skeptical scrutiny of all hypotheses that are served up to us and at the same time a great openness to new ideas. If you are only skeptical, then no new ideas make it through to you. On the other hand, if you are open to the point of gullibility and have not an ounce of skeptical sense in you, then you cannot distinguish useful ideas from the worthless ones.

Why, we might inquire, do some people prefer orthodoxy while others favor heresy? Is there a personality trait—a temperament tendency, as it were—for tradition, and another for change? This is an important question because the answer helps explain why in the history of science some chose to support radical new ideas while others opposed them, and who today (or even tomorrow) will choose to support the status quo versus the cutting edge.

In 1990 David Swift published a book titled *SETI Pioneers: Scientists*

Talk About Their Search for Extraterrestrial Intelligence, in which he identified an overabundance of firstborns, including Sagan. But is it a statistically significant overabundance? Swift did not test for this, but UC Berkeley psychologist Frank J. Sulloway and I did. For the SETI pioneer group eight is the expected number of firstborns based on the number of siblings they had, but twelve is the observed number. This difference is statistically significant at the 95 percent level of confidence.

What does this finding mean? In Sulloway's 1996 book *Born to Rebel* he presents a summary of 196 controlled birth-order findings classified according to what is known as the Five Factor Model of personality:

Conscientiousness: Firstborns are more responsible, achievement-oriented, and planful.

Agreeableness: Laterborns are more easygoing, cooperative, and popular.

Openness to Experience: Firstborns are more conforming, traditional, and closely identified with parents.

Extroversion: Firstborns are more extroverted, assertive, and likely to exhibit leadership.

Neuroticism: Firstborns are more jealous, anxious, and fearful.

To measure Sagan's personality Sulloway and I requested a number of his family members, friends, and colleagues to rate him on a standardized personality inventory of forty descriptive adjectives using a nine-step scale. For example: *I See Carl Sagan as Someone Who Was . . .* hardworking or lackadaisical, tough-minded or tender-minded, assertive or unassertive, organized or disorganized, rebellious or conforming, etc. The following results are in percentile rankings relative to Sulloway's database of more than 7,276 subjects.

Most consistent with his firstborn status was Sagan's exceptionally high ranking—88th percentile—on conscientiousness (ambitiousness, dutifulness) and his strikingly low ranking of the 13th percentile on agreeableness (tender-mindedness, modesty). But his openness to experience (preference for novelty) was nearly off the scale, at the 97th percentile. Why? First, birth order is hardly the only influence on openness and can be affected by cultural influences and social attitudes—Sagan was raised in a socially liberal Jewish family, and he was mentored by such scientific revolutionaries as Joshua Lederberg and H. J. Muller. Second, openness also includes an "intellectual" component, and firstborns tend to excel at intellectual pursuits, as reflected by their higher IQ scores and a tendency to win more Nobel Prizes in science.

Here is the key to understanding the exquisite balance between tradition and change: Sagan's high openness led him to be a SETI pioneer, but his high conscientiousness made him skeptical of UFOs. In fact, Sagan's openness led him to gamble on a number of radical ideas, but his conscientiousness prevented him from taking them too far, into crankdom. His low agreeableness meant he did not suffer fools gladly, which can be a virtuous trait in the skeptical business.

Complementing this analysis, Sulloway and I distributed the same survey to eight of Stephen Jay Gould's friends and colleagues, with almost identical results (that included a 0.92 inter-rater reliability score).

Sagan and Gould were not only close friends and colleagues, they also shared many of the same skeptical attitudes and tendencies (not to mention family background and cultural upbringing), leading us to hypothesize that one key to finding the exquisite balance is being exceptionally open-minded while simultaneously maintaining high conscientiousness and tough-mindedness, which is a standard indicator of low agreeableness.

No one exhibits a personality trait all the time, of course. These traits are variable tendencies, but as such we can understand why some people tend toward orthodoxy while others tend toward heresy. The balance between the two is what allows science to operate so effectively, and science has demonstrated how and why this should be so. How recursive!

17

The Enchanted Glass

Francis Bacon and experimental psychologists show why the facts in science never just speak for themselves

In the first trimester of the gestation of science in the early modern period, one of its midwives, Francis Bacon, penned an immodest work titled *Novum Organum* ("new tool," after Aristotle's *Organon*), which would open the gates to the "Great Instauration" he hoped to inaugurate through the scientific method. Rejecting both the unempirical tradition of scholasticism and the Renaissance quest to recover and preserve ancient wisdom, Bacon sought a blend of sensory data and reasoned theory.

A major impediment to his goal of reconceptualizing all natural knowledge, however, was psychological barriers that colored clear judgment of the facts, of which Bacon identified four types: *idols of the cave* (individual peculiarities), *idols of the marketplace* (limits of language), *idols of the theater* (preexisting beliefs), and *idols of the tribe* (inherited foibles of human thought): "Idols are the profoundest fallacies of the mind of man. Nor do they deceive in particulars . . . but from a corrupt and crookedly-set predisposition of the mind; which doth, as it were, wrest and infect all the anticipations of the understanding."

Recently experimental psychologists have corroborated Bacon's idols, particularly those of the tribe, in the form of numerous cognitive biases. The self-serving bias, for example, dictates that we tend to see ourselves in a more positive light than others see us: national surveys show that most businesspeople believe they are more moral than other businesspeople, while psychologists who study moral intuition think they are more moral than other such psychologists. In one College Entrance Examination Board survey of 829,000 high school seniors, 0 percent rated themselves below average in "ability to get along with others," while 60 percent put themselves in the top 10 percent (presumably not all were from Lake Wobegon). And according to a 1997 *U.S. News & World Report* study on who Americans believe are most likely to go to heaven, 52 percent said Bill Clinton; 60 percent thought Princess Diana; 65 percent chose Michael Jordan; 79 percent selected Mother Teresa; and, at 87 percent, the person most likely to go to heaven was the survey taker!

Princeton University psychology professor Emily Pronin and her colleagues tested an idol called "bias blind spot," in which subjects recognized the existence and influence in others of eight different cognitive biases, but they failed to see those same biases in themselves. In one study on Stanford University students, when asked to compare themselves to their peers on such personal qualities as friendliness and selfishness, they predictably rated themselves higher. Even when the subjects were warned about the "better than average" bias and asked to reevaluate their original assessments, 63 percent claimed that their initial evaluations were objective, and 13 percent even claimed to be too modest! In a second study, Pronin randomly assigned subjects high or low scores on a "social intelligence" test. Unsurprisingly, those given the high marks rated the test fairer and more useful than those receiving low marks. When asked if it was possible that they had been influenced by the score on the test, subjects responded that other participants had been far more biased than they were. In a third study, in which Pronin queried subjects about what method they used to assess their own and others' biases, she found that people tend to use general theories of behavior when evaluating others, but use introspection when appraising themselves; but in what is called the "introspection illusion," people do not believe that others can be trusted to do the same. Okay for me but not for thee. Here is how she described it to me in an email:

> Also, one of the ideas which you may be interested in involves something I and my colleagues, Lee Ross and Thomas Gilovich, have called the

> "introspection illusion." This refers to the following tendency: We view our perceptions of our mental contents and processes as the gold standard for understanding our actions, motives, and preferences. But, we do not view others' perceptions of their mental contents and processes as the gold standard for understanding their actions, motives, and preferences. This "illusion" that our introspections are a gold standard leads us to introspect to find evidence of bias and we are thus likely to infer that we have not been biased—since most biases operate outside of conscious awareness.

Overall, Pronin has found, even when subjects admit to having a bias, such as being a member of a partisan group, this "is apt to be accompanied by the insistence that, in their own case, this status . . . has been uniquely *enlightening*—indeed, that it is the *lack* of such enlightenment that is making those on the other side of the issue take their misguided position."

Psychologist Frank Sulloway at the University of California at Berkeley and I made another discovery of an idol-like distortion called *attribution bias* in a study we conducted on why people say they believe in God, and why they think other people believe in God. In general, most people attribute their own belief in God to such intellectual reasons as the good design and complexity of the world, whereas they attribute others' belief in God to such emotional reasons as it is comforting, gives meaning, and that they were raised to believe.

None of these findings would surprise Francis Bacon who, four centuries ago, noted: "For the mind of man is far from the nature of a clear and equal glass, wherein the beams of things should reflect according to their true incidence; nay, it is rather like an enchanted glass, full of superstition and imposture, if it be not delivered and reduced."

18

Fahrenheit 2777

9/11 has generated the mother of all conspiracy theories

When the noted French left-wing activist Thierry Meyssan's 9/11 conspiracy book *L'Effroyable Imposture* became a best seller in 2002, I never imagined an "appalling deception" such as this would ever find a voice in America. But at a recent public lecture I was buttonholed by a Michael Moore–wannabe filmmaker who breathlessly explained that 9/11 was orchestrated by Bush, Cheney, Rumsfeld, and the CIA to implement their plan for global domination and a New World Order, to be financed by G.O.D. (Gold, Oil, Drugs) and launched by a Pearl Harbor–like attack on the World Trade Center and the Pentagon, thereby providing the justification for war. The evidence is there in the details, he explained, handing me a faux dollar bill (9-11 replacing 1, Bush supplanting Washington) chockablock full of web sites.

In fact, if you type into Google "World Trade Center conspiracy" you'll find 643,000 sites, where you will discover that the Pentagon was hit by a missile, air force jets were ordered to "stand down" and not intercept Flights 11 and 175, which struck the twin towers, which themselves were razed by demolition explosives timed to go off soon after the impact of the planes, a mysterious white jet shot down Flight 93 over Pennsylvania, New York Jews were ordered to stay home that day (Zionists and other pro-Israeli factions, of course, were involved), and much more in what one site calls the "9/11 Truth Movement." Books also abound, including *Inside Job* by Jim Marrs, *The New Pearl Harbor* by David Ray Griffin, and *9/11: The Great Illusion* by George Humphrey. The single best debunking of this conspiratorial codswallop is in the March 2005 issue of *Popular Mechanics*, an exhaustive point-by-point analysis of the most prevalent claims.

The belief that a handful of unexplained anomalies can undermine a well-established theory lies at the heart of all conspiratorial thinking (as well as creationism, Holocaust denial, and crank theories of physics), and is easily refuted by noting that beliefs and theories are not built on single facts alone, but on a convergence of evidence from multiple lines of inquiry. All of the "evidence" for a 9/11 conspiracy falls under the rubric of this fallacy.

For example, according to 911research.wtc7.net, steel melts at a temperature of 2,777 degrees Fahrenheit, but jet fuel burns at only 1,517 degrees F. No melted steel, no collapsed towers. "The planes did not bring those towers down; bombs did," says abovetopsecret.com. Wrong. In an article in the *Journal of the Minerals, Metals, and Materials Society*, MIT engineering professor Dr. Thomas Eager explains why: steel loses 50 percent of its strength at 1,200 degrees F; 90,000 liters of jet fuel ignited other combustible materials such as rugs, curtains, furniture, and paper, which continued burning after the jet fuel was exhausted, raising temperatures above 1,400 degrees F and spreading the fire throughout the building; temperature differentials of hundreds of degrees across single steel horizontal trusses caused them to sag, straining and then breaking the angle clips that held them to the vertical columns; once one truss failed, others failed, and when one floor collapsed (along with the ten stories above it) onto the next floor below, that floor then gave way, creating a pancaking effect that triggered the 500,000-ton building to collapse. Conspiricists argue that the buildings should have fallen over on their sides, but with 95 percent of each building consisting of air, they could only have collapsed straight down.

All of the 9/11 conspiracy claims are this easily refuted. On the Pentagon "missile strike," for example, I queried my antagonist about what happened to Flight 77, which disappeared at the same time. "The plane was destroyed and the passengers were murdered by Bush operatives," he solemnly revealed. "Do you mean to tell me that not *one* of the thousands of conspirators needed to pull all this off," I retorted, "is a whistle-blower who would go on TV or write a tell-all book?" My rejoinder was met with the same grim response I get from UFOlogists when I ask them for concrete evidence: men in black silence witnesses, and dead men tell no tales.

To call all this *bullshit* risks denigrating a fine descriptive adjective to the mere excremental.

Update: The 9/11 "truthers" are just as active today as when I originally wrote this essay, and we (and others) continue to rebut their claims. On controlled demolitions, for example, see the Skeptic *magazine article here:* http://bit.ly/lu00vbL

III

PSEUDOSCIENCE AND QUACKERY

19

Smart People Believe Weird Things

Rarely does anyone weigh facts before deciding what to believe

When men wish to construct or support a theory, how they torture facts into their service!

—Charles Mackay, *Extraordinary Popular Delusions and the Madness of Crowds*, 1852

In April 1998, when I was on a lecture tour for my book *Why People Believe Weird Things*, the psychologist Robert Sternberg attended my presentation at Yale. His response to the lecture was both enlightening and troubling. It is certainly entertaining to hear about other people's weird beliefs, Sternberg reflected, because we are confident that we would never be so foolish as to believe in such paranormal ephemera as aliens, astrology, ESP, ghosts, and the like. But why do *smart* people believe weird things? Sternberg's challenge led to a second edition of my book, with a new chapter expounding on my answer to his question: *smart people believe weird things because they are skilled at defending beliefs they arrived at for non-smart reasons.*

Most of us most of the time come to our beliefs for a variety of reasons having little to do with empirical evidence and logical reasoning (that, presumably, smart people are better at employing). Rather, such variables as genetic predispositions, parental predilections, sibling influences, peer pressures, educational experiences, and life impressions all shape the personality preferences and emotional inclinations that, in conjunction with numerous social and cultural influences, lead us to make certain belief choices. Rarely do any of us sit down before a table of facts, weigh them pro and con, and choose the most logical and rational belief, regardless of what we previously believed. Instead, the facts of the world come to us through the colored filters of the theories, hypotheses, hunches, biases, and prejudices we have accumulated through our lifetime. We then sort through the body of data and select those most confirming what we already believe, and ignore or rationalize away those that are disconfirming.

This phenomenon, called the *confirmation bias*, helps explain the findings published in the National Science Foundation's biennial report (April 2002) on the state of science understanding, that 30 percent of adult Americans believe that UFOs are space vehicles from other civilizations, 60 percent believe in ESP, 40 percent think that astrology is scientific, 32 percent believe in lucky numbers, 70 percent accept magnetic therapy as scientific, and 88 percent agree that alternative medicine is a viable means of treating illness.

Education by itself is no paranormal prophylactic. Although belief in ESP decreased over a decade from 65 percent among high school graduates to 60 percent among college graduates, and belief in magnetic therapy dropped from 71 percent among high school graduates to 55 percent among college graduates, that still leaves more than half fully endorsing such claims! And for embracing alternative medicine, the percentages actually *increase*, from 89 percent for high school grads to 92 percent for college grads.

We can glean a deeper cause to this problem in another statistic: 70 percent of Americans still do not understand the scientific process, defined in the study as grasping probability, the experimental method, and hypothesis testing. One solution is more and better science education, as indicated by the fact that 53 percent of Americans with a high level of science education (nine or more high school and college science/math courses) understand the scientific process, compared to 38 percent with a middle level (six to eight such courses) science education, and 17 percent with a low level (less than five such courses).

The key here is teaching how science works, not just what science has discovered. In 2002 we published an article in *Skeptic* (vol. 9, no. 3) revealing the results of a study that found no correlation between science knowledge (facts about the world) and paranormal beliefs. The authors,

W. Richard Walker, Steven J. Hoekstra, and Rodney J. Vogl, concluded: "Students that scored well on these [science knowledge] tests were no more or less skeptical of pseudoscientific claims than students that scored very poorly. Apparently, the students were not able to apply their scientific knowledge to evaluate these pseudoscientific claims. We suggest that this inability stems in part from the way that science is traditionally presented to students: Students are taught *what* to think but not *how* to think."

To attenuate these paranormal belief statistics we need to teach people (everyone, not just students) that science is not a database of unconnected factoids, but a set of methods designed to describe and interpret phenomena, past or present, aimed at building a testable body of knowledge open to rejection or confirmation. Science is a way of thinking that recognizes the need to test hypotheses so that the process is not reduced to mere opinion mongering, that the findings of such tests are provisional and probabilistic, and that natural explanations are always sought for natural phenomena.

Lacking a fundamental comprehension of how science works, the siren song of pseudoscience becomes too alluring to resist, no matter how smart you are.

20

Mesmerized by Magnetism

An eighteenth-century investigation into mesmerism shows us how to think about twenty-first-century therapeutic magnets

In an August 11, 1997, ABC *World News Tonight* report on "biomagnetic therapy," a physical therapist explained that "magnets are another form of electric energy that we now think has a powerful effect on bodies." A fellow selling $89 magnets proclaimed: "All humans are magnetic. Every cell has a positive and negative side to it."

On the positive side, these magnets are so weak that they cause no harm. On the negative side, these magnets do have the remarkable power of attracting the pocketbooks of gullible Americans to the tune of about $300 million a year. The magnets range in size from coin-size patches to king-size mattresses, and their curative powers are said to be nearly limitless, based on the premise that magnetic fields increase blood circulation and enrich oxygen supplies because of the iron in the blood.

This is fantastic flapdoodle and a financial flimflam. Iron atoms in a magnet are crammed together in a solid state about one atom apart from each other. In your blood only four iron atoms are allocated to each hemoglobin molecule, and they are separated by distances too great to form a magnet. This is easily tested by pricking your finger and placing a drop of your blood next to a magnet.

In addition, the commercially sold therapeutic magnets are too weak for their field to penetrate the skin. Try this experiment: place a dozen sheets of paper between one of these magnets and your refrigerator door. The magnet will drop faster than Enron stock. For my now-defunct Family Channel television series *Exploring the Unknown*, we placed two real magnets and two fake magnets on the backs of subjects and then measured their skin temperature with an infrared camera. There were no differences in temperatures, which means there were no differences in circulation between the fake and the real magnets.

But what about claims that magnets attenuate pain? In a Baylor College of Medicine double-blind study of fifty patients (where twenty-nine got real magnets and twenty-one sham magnets), 76 percent in the experimental group but only 19 percent in the control group reported a reduction in pain. Unfortunately this study included only one forty-five-minute treatment, did not try other pain-reduction modalities, did not record the length of the pain reduction, and was never replicated.

Scientists studying magnetic therapy would do well to read the 1784 "Report of the Commissioners Charged by the King to Examine Animal Magnetism" (reprinted in an English translation in *Skeptic*, vol. 4, no. 3), instituted by King Louis XVI of France and conducted by Benjamin Franklin and Antoine Lavoisier to test the claims of the German physician Franz Anton Mesmer, discoverer of "animal magnetism." Mesmer reasoned that just as an invisible force of gravity binds the planets together, and an invisible force of magnetism draws iron shavings to a lodestone, so an invisible force—animal magnetism—flows through living beings, blockage of which can cause disease. Cure comes through releasing the blockage.

The experimenters began by trying to magnetize themselves, to no effect. They then tried seven people from the lower classes and compared their results against seven people from the upper classes. Only three, all from the lower classes, experienced anything significant, so the commission concluded it was due to the power of suggestion. To test the null hypothesis that magnetism was all in the mind, Franklin and Lavoisier deceived some subjects into thinking they were receiving the experimental treatment (magnetism) when they really were not, while others did receive the treatment and were told that they had not. The results were clear: the effects were due to the power of suggestion only.

In a second experiment Franklin had Mesmer's representative Charles d'Eslon magnetize a tree in his garden. The subject walked around the garden hugging trees until he collapsed in a fit in front of the fourth tree; it was the fifth one that was "magnetized." In another test a woman was blindfolded and told that d'Eslon was "influencing" her (he wasn't), causing her to collapse in a mesmeric "crisis." Another woman could supposedly sense "magnetized" water. Lavoisier filled several cups with water, only one of which was "magnetized." After touching an unmagnetized cup she collapsed in a fit, upon which Lavoisier gave her the "magnetized" one, which "she drank quietly & said she felt relieved."

The commission concluded that "nothing proves the existence of Animal-magnetism fluid; that this fluid with no existence is therefore without utility; that the violent effects observed at the group treatment belong to touching, to the imagination set in action & to this involuntary imitation that brings us in spite of ourselves to repeat that which strikes our senses." In other words, the effect is mental, not magnetic.

The control of intervening variables and the testing of specific claims, without resort to unnecessary hypothesizing about what is behind the "power," are the lessons modern skeptics should take from this historical masterpiece, as well as the sad fact that true believers remain unaffected by contradictory evidence, in the eighteenth century as well as today.

21

Show Me the Body

Purported sightings of Bigfoot, Nessie, and Ogopogo fire our imaginations. But anecdotes alone do not make a science

In 1895 the French novelist Anatole France opined, "Chance is perhaps God's pseudonym when he does not want to sign." Perhaps, but as another observer of matters human noted, "sometimes a cigar is just a cigar." Thus, although they say that celebrities pass on in clusters of three, it is surely coincidence that in January 2003 the world lost the creators of two of its most celebrated biohoaxes in Douglas Herrick, father of the risibly ridiculous "jackalope" (half jackrabbit, half antelope), and Ray L. Wallace, paternal guardian of the less absurd and more widely believed Bigfoot.

The jackalope enjoins laughter in response to such peripheral hokum as hunting licenses sold only to those whose IQs range between 50 and 72, bottles of the rare but rich jackalope milk, and additional evolutionary hybrids such as the jackapanda. Bigfoot, on the other hand, while occasionally eliciting an acerbic snicker, enjoys greater plausibility for a simple evolutionary reason: large hirsute apes presently roam the forests of Africa, and at least one species of a giant ape—*Gigantopithecus*—flourished several hundred thousand years ago alongside our hominid ancestors.

Is it possible that a real Bigfoot lives despite the posthumous confession by the Wallace family that it was all just a practical joke by a fun-loving prankster? Certainly. After all, although Bigfoot proponents do not dispute the evidence that Wallace tromped around in strap-on Shaq-o-size wooden feet, they correctly note that tales of the giant yeti living in the Himalayas, and Native American lore about Sasquatch wandering about the Pacific Northwest, all emerged long before Wallace pulled his prank in 1958.

In point of fact, throughout much of the twentieth century it was entirely reasonable to speculate about and search for Bigfoot, as it was the creatures of Loch Ness, Lake Champlain, and Lake Okanagan (Nessie, Champ, and Ogopogo, respectively), and even extraterrestrials. Science traffics in the soluble, so while jackalopes do not warrant our limited exploratory resources, for a time these other creatures did. Why don't they now?

The study of animals whose existence has yet to be proven is known as cryptozoology, a term coined in the late 1950s by the Belgian zoologist Bernard Heuvelmans. Cryptids, or "hidden animals," begin life as muddy footprints, blurry photographs, grainy videos, and countless anecdotes about strange things that go bump in the night. Cryptids come in many forms, including the aforementioned giant pongidid and lake monsters, as well as sea serpents, giant octopi snakes, and birds, and even living dinosaurs (the most famous being *mokele-mbembe*, purportedly swamping through lakes of West Africa).

The reason why cryptids merit our attention is that there have been enough successful discoveries made by scientists based on local anecdotes and folklore that we cannot dismiss all claims a priori. The most famous examples include the gorilla in 1847 (and the mountain gorilla in 1902), the giant panda in 1869, the okapi (a short-necked relative of the giraffe) in 1901, the Komodo dragon in 1912, the bonobo (or pygmy chimpanzee) in 1929, the megamouth shark in 1976, the giant gecko in 1984, the beaked whale in 1991, and the Spindlehorn ox from Vietnam in 1992. Cryptozoologists are especially proud of the catch in 1938 of a coelacanth, an archaic-looking species of fish believed by zoologists to have gone extinct in the Cretaceous Period.

Although discoveries of new species of bugs and bacteria are routinely published in the annals of biology, these examples are startling because of their recency, size, and commonality to the aforementioned famous cryptid cousins Big Foot, Nessie, et al. But please note what all these examples have in common—a body! To name a new species one must have a type specimen—a holotype—from which a detailed description can be made, photographs taken, models cast, and a professional scientific analysis published.

Anecdotes are a good place to begin an investigation, but anecdotes by themselves do not constitute a new species. In fact, in the words of the UC Berkeley social scientist Frank Sulloway—words that should be elevated to a maxim: "Anecdotes do not make a science. Ten anecdotes are no better than one, and a hundred anecdotes are no better than ten."

I employ Sulloway's maxim every time I encounter Bigfoot hunters, Loch Ness seekers, or alien abductees. The anecdotal tales make for gripping narratives, but they do not make for sound science. After a century of searching for these chimerical creatures, until a body is produced skepticism is the appropriate response.

22

What's the Harm?

Alternative medicine is not everything to gain and nothing to lose

After being poked, prodded, scanned, drugged, and radiated your doc tells you that there is nothing more that can be done to cure what ails you. Why not try an alternative healing modality? What's the harm?

I started thinking about this question in 1991 when my normally intelligent mother presented to a psychiatrist symptoms of cognitive confusion, emotional instability, and memory loss. Within an hour it was determined that she was depressed. I didn't buy it. My mom was weird, not depressed. I requested a second opinion, from a neurologist. A minor deception got us in that afternoon (I'm a PhD, not an MD).

A CT scan revealed an orange-size meningioma tumor. Meningioma tumors (originating on the meninges, the protective lining of the brain) are far more common in women than men—the ratio ranges from 1.4:1 to 2.8:1—and cluster, curiously, in Los Angeles County, where we live. They are usually successfully treated with surgical removal. Indeed, within days my mom was back to her bright and cheery self—what a remarkably recuperative and pliable organ is the brain. Unfortunately, within a year my mom had two new tumors in her brain. (My mom might be another data point in support of Harvard Medical School surgeon Judah Folkman's angiogenesis theory of cancer—that tumors secrete chemical stimulants that draw blood vessels to them for growth, as well as chemical inhibitors that prevent other tumors from implementing their own angiogenesis programs. Remove the dominant tumor, however, and you eliminate the angiogenesis inhibitors, thus allowing the dormant tumors to spring to life.) Three more rounds of this cycle of surgical removal and tumor return, plus two doses of gamma-knife radiation (a pinpoint-accurate beam that destroys cancer cells), finally led to the dreaded prognosis: there was nothing more to be done.

What is a skeptic to do? An ideological commitment to science is one thing, but this was my mom! I turned to the literature, and with the help of our brilliant and humane oncologist, Dr. Avrum Bluming, determined that we would try an experimental treatment, mifepristone, a synthetic antiprogestin better known as RU-486, which is also an induced abortion drug. A small-sample study suggested that it might retard the growth of tumors. It didn't. My mom was dying. There was nothing to lose in trying some alternative cancer treatments, right? Wrong.

The choice is not between scientific medicine that doesn't work and alternative medicine that might work. Instead, there is only scientific medicine that has been tested and everything else ("alternative" and "complementary" medicine) that has not been tested. A few reliable authorities test and review the evidence for some of the claims—notably Dr. Stephen Barrett's Quackwatch (www.quackwatch.org), Dr. William Jarvis's National Council Against Health Fraud (www.ncahf.org), and Dr. Wallace Sampson's journal *The Scientific Review of Alternative Medicine.*

However, most alternatives slip under the scientific peer review radar. This is why it is alarming that, according to the American Medical Association, the number of visits to alternative practitioners exceeds visits to traditional medical doctors; the amount of money spent on herbal medicines and nutrition therapy accounts for more than half of all out-of-pocket expenses to physicians; and most disturbingly, 60 percent of patients who underwent alternative treatments did not report that information to their physician—a serious, and even potentially fatal, problem if herbs and medicines are inappropriately mixed.

For example, the September 17, 2003, issue of the *Journal of the American Medical Association* reported the results of a study that found St. John's wort, an herb derived from a blooming *Hypericum perforatum* plant and hugely popular as an alternative elixir (to the tune of $86 million in 2002), can significantly impair the effectiveness of dozens of medications, including those used to treat high blood pressure, cardiac arrhythmias, high cholesterol, cancer, pain, and depression. The study's authors show that St. John's wort affects the liver enzyme cytochrome P450 3A4, essential to metabolizing at least half of all prescription drugs, speeding up the breakdown process, and thereby shortchanging patients of their lifesaving medications.

But there is a deeper problem. All of us are limited to a few score years in which to enjoy meaningful life and love. Time is precious and fleeting. Given the choice of spending the next couple of months schlepping my dying mother around the country on a wild goose chase versus spending quality time together, my dad and I opted for the latter. She died a few months later, on September 2, 2000, three years from the day I penned the column on which this essay is based.

Medicine is miraculous and science is scintillating, but in the end life ultimately turns on the love of the people who matter most. It is for those relationships, especially, that we should apply the ancient medical principle *primum non nocere*—first do no harm.

23

Bunkum!

Broad-mindedness is a virtue when investigating extraordinary claims, but often they turn out to be pure bunk

Those of us who practice skepticism for a living often find ourselves tiptoeing politely around the PC police who believe that truth is relative and all opinions are to be respected. Thus, when asked "Are you a debunker?" my initial instinct is to dissemble and mutter something about being an investigator, as if that will soften the blow.

But what need, really, is there to assuage? According to the *Oxford English Dictionary*, to "debunk" is to "remove the nonsense from; to expose false claims or pretensions." "Bunk" is slang for "humbug," and "bunkum" is "empty claptrap oratory." Here is some bunk that merits no brook.

Aliens did not crash in Roswell, New Mexico, or anywhere else. If there are aliens in the cosmos they very likely are too far away to have made it to one of a hundred billion stars in one of a hundred billion galaxies. Alien abductees are not visiting the mother ship; they are having nightmares and wet dreams, or creating false memories through hypnosis conducted by abduction "therapists."

JFK was shot by Lee Harvey Oswald, not by a KGB, CIA, FBI cabal in cahoots with the military-industrial complex, Mafia, and Castro. Neither George Bush Sr. nor Jr. are puppets of the Freemasons, Illuminati, Rockefellers, Rothchilds, or any other New World Order secret society.

There was, however, a conspiracy to land a man on the moon, and it succeeded.

Feng shui—the ancient Chinese tradition (recently imported to the West) of arranging furniture, doors, windows, and various objects so as to adjust the yin-yang energy flow through a home or building to bring about health, harmony, and kismet—has nothing to do with mystical forces and everything to do with Chinese geography. Doors and windows do not control the movement of "ch'i energy" (or "life force") because there is no such thing, but they do regulate the flow of cold wind blowing off a mountain, regardless of whether the mountain resembles a dragon, snake, or tiger. A bed placed in front of a door does not block ch'i, but it does interfere with the design aesthetics. Call an interior designer, not a feng shui practitioner.

Ear coning cleans your ears and mind. Lie down on your side with your head on a pillow. Place a long, narrow, cylindrical cone of wax into your ear canal until there is a tight seal. Light the open end of the cone on fire. The negative pressure created will not only remove undesirable ear wax, according to Coning Works in Sedona, Arizona; additional benefits include "spiritual opening and emotional clearing, realignment and cleansing of subtle energy flows, sharpening of mental functioning, vision, hearing, smell, taste and color perception" and, most importantly, it "acts as a catalyst to clear out debris from nerve endings allowing for clear vibrational flow to corresponding areas of mind, body and spirit." Why pay $25 to $75 to have your ears cleaned by your doctor, asks Wholistic Health Solutions, "when you can easily do it at home?"

Well, for starters, according to a 1996 study conducted by physicians at the Spokane, WA, Ear, Nose, and Throat Clinic and published in the journal *Laryngoscope*, "Tympanometric measurements in an ear canal model demonstrated that ear candles do not produce negative pressure," and thus there was no removal of wax in the eight ears tested. Worse, a survey of 122 otolaryngologists (ear, nose, and throat docs) identified twenty-one ear injuries from ear coning ("which end am I supposed to light?"). If one is inclined toward such self-mutilation (or a good chortle), however, I recommend a quick stop at buttcandle.com, where you can find a "gentler alternative to laxatives, enemas and anti-flatulence pills" in the form of a carefully (and gently) placed hollow candle that when burning creates a vacuum that draws out impurities. Best of all, it's "100% soluble and septic-safe."

Laundry balls clean clothes. Putting into a wash spherical, toroidal, or spiked balls that contain no chemicals and are indefinitely reusable to clean, deodorize, sterilize, bleach, and soften clothes, do not "ionize," "structure," "cluster," or "magnetize" water, as various manufacturers claim. The EarthSmart Laundry CD, Unbelievable Laundry Disk 2000, WashaBall, EnviWash Laundry Ball, Bion Ceramic Laundry Disk, Scrub Balls, EuroWash Laundry Ball, Natural Wash Laundry Balls, LaundryMaster Ionic LaundryBall, Turbo Plus Laundry Disc, Stereo Laundry Disc, Little Helper Laundry Balls, Dynamic One Laundry Clean Ring, ABI Laundry Ball, CleanTec Washing Stones, CW-6 Laundry Ball, and EcoSave Magnetic Washball all work on the same principle: washing clothes in soapless warm water does have some cleansing effect, particularly for nongreasy clothes mainly soiled by dust, dirt, and sweat. But with laundry balls costing from $25 to $75, golf balls are just as effective and a lot cheaper.

A counterfeit pen can detect counterfeit bills. Containing tincture of iodine that reacts with the starch in recycled paper to create a black streak, the pen only works to catch counterfeiters brainless enough to use cheap paper, thus creating a false sense of security. Meanwhile, clever counterfeiters with brains who use high-quality fiber or linen paper containing no starch or whitening continue to fleece their marks. Merchants beware: after warning law enforcement agencies—who ignored him—fellow skeptic James Randi periodically applies commercial spray starch on $50 and $100 bills for recirculation into the economy in the hopes that false pen positives will force the bunkum squads into action.

To "buncomize" is to "talk bunkum," and no one does this with a better vocabulary than pseudoscientists, who lace their hokum narratives with scientistic jargon. (One laundry ball manufacturer claims that it "works on 'Quantum Mechanics' [Physics], not chemistry, with a method called 'Structured Water Technology.'" Another uses "infra-red waves that change the molecular structure of the water.") To "do a bunk" is to "make an escape" or "to depart hurriedly," a wise move when skeptics arrive on the scene fully armed with steel-jacketed science and armor-piercing reason.

24

Magic Water and Mencken's Maxim

Social critic H. L. Mencken offers a lesson on how to respond to outrageous pseudoscientific claims

Henry Louis Mencken was a stogie-chomping, QWERTY-pounding, social commentator in the first half of the twentieth century who never met a man or a claim he didn't like . . . to disparage, critique, or parody with wit that would shame Dennis Miller back to *Monday Night Football.* Covering the Scopes "Monkey Trial" for the *Baltimore Sun* in 1925, for example, of three-time presidential candidate and defender of the faith, William Jennings Bryan, Mencken wrote: "Once he had one leg in the White House and the nation trembled under his roars. Now he is a tin-pot pope in the Coca-Cola belt and a brother to the forlorn pastors who belabor half-wits in galvanized iron tabernacles behind the railroad yards."

Stupidity and quackery were favorite targets for Mencken's barbs. "Nature abhors a moron," he once quipped. "No one in this world, so far as I know . . . has ever lost money by underestimating the intelligence of the great masses of the plain people," he famously noted. Some claims are so preposterous, in fact, that there is only one rejoinder: "One horse-laugh is worth ten-thousand syllogisms." I call this "Mencken's Maxim," and I find that it is an appropriate response to preposterous claims made about improved water sold on the web. Dozens of sites sell this crystal-clear snake oil—and some even take American Express. I offer as a holotype of

Mencken's Maxim the following: Golden "C" Lithium Structured Water (http://www.luminanti.com/goldenc.html).

This "is pure water infused with the energies of the Golden 'C' crystal, a very special and extremely rare stone mined near San Diego at the turn of the 20th century." The stone, continuing, "contains more lithium than any other stone on the planet. It also contains the rare signature secret element, gallium. Gallium, in combination with the high lithium content as well as beryllium and the family of high frequency crystals such as emeralds and aquamarine; Golden 'C' emits a signature one-of-a-kind healing energy." How does the Golden "C" water get these magical qualities? Crystal and water are placed in a ceramic container in a "dark and quiet space" for twenty-four hours, then the water is poured into "violet glass bottles" that "energize it." Finally, "each violet bottle is placed precisely within a special copper pyramid, specially designed to have the exact Sacred Geometry to create a Pillar of Light Jacob's Ladder vortex." A personal "vibrational" touch comes through, "adding your special crystal to the diluted water," or by placing "the stem end of an activated tuning fork to the Golden 'C' bottle."

At only $15 per half ounce, Golden "C" water is a bargain because it "aligns and balances chakras and meridians; acts as a negative ion generator; clears stressful emotions and negative thought forms; clears all negative energy from crystals, food, rooms, people, and pets; eases stress; disperses anger; improves [the] immune system; clears [the] bed of nightmare energy and previous energy of dreams; improves mental concentration; facilitates deeper meditations; hydrates and soothes skin; creates environment for visionary dreams." And, most importantly, it "clears and protects from electromagnetic pollution such as kitchen appliances, TV, microwave emissions from ovens and the environment, electrical clocks, stereos, high electrical wire lines, etc." As evidence we are offered this factoid: "Using an instrument to measure wavelengths of light, Holy Water from Lourdes, France registered 156,000 angstroms of light. Golden 'C' water registered 250,000 angstroms of light!"

Wait! That's not Mencken's moment. Just below the order button a warning label reads: "Note: no actual lithium is in the water. Only the energetics of lithium and the other minerals is contained in the water." Maybe that explains another disclaimer, perhaps written with attorneys in mind: "No therapeutic, drug or healing claims related to the physical body are made in the use of Golden 'C' Lithium Structured Water." However, one is advised to keep it refrigerated.

In case any credulity remains, according to Ray Beiersdorfer, professor of geochemistry at Youngstown State University in Ohio, "exposing ordinary water to lithium crystals, or any other crystals for that matter, cannot fundamentally alter the molecular structure of the water. The chemical structure within the water molecule, as defined by bond length and orientation, doesn't change. The claim that the chemical structure of liquid water changes due to exposure to a relatively insoluble crystal is nonsense."

For another Mencken moment check out tachyonized superconductor water at www.tachyon-energy-products.com. Its promoter, Gene Latimer, explains its benefits: "I am now living in a radically different electromagnetic field environment that appears to be harmonizing the chaotic impact of electrical Alternating Current on the life forms in our house." *All* life forms? Wow! And guess what? Tachyon is not limited to water. You can order tachyonized gel, algae, spirulina, herbs, mattress pads, massage oil, a cell phone disk, and even "star dust." Sprinkle lightly.

Considering these sites, we would all do well to follow another Mencken maxim: "I believe it is better to tell the truth than to lie. . . . And I believe that it is better to know than to be ignorant." Amen, brother.

25

Death by Theory

Attachment therapy is based on a pseudoscientific theory that, when put into practice, can be deadly

In April 2000, ten-year-old Candace Newmaker began treatment for "Attachment Disorder" (AD). Her adoptive mother of four years, Jeane Newmaker, was having trouble handling what she considered to be Candace's disciplinary problems, and when she sought help from a therapist affiliated with the Association for Treatment and Training in the Attachment of Children (www.ATTACh.org), was told Candace needed Attachment Therapy (AT), based on the theory that if a normal attachment is not formed during the critical first two years, then reattachment can be done later.

According to the theory, first the child must be subjected to physical "confrontation" and "restraint" to release repressed abandonment anger. The process repeats for as long as is necessary—day after day, week after week—until the child is physically exhausted and emotionally reduced to an "infantile" state. Then the parents cradle, rock, and bottle-feed the child, implementing a "reattachment."

Candace was taken to Evergreen, Colorado, where she was treated by Connell Watkins, a nationally prominent attachment therapist and past clinical director for the Attachment Center at Evergreen (ACE), and her associate Julie Ponder, a recently licensed family counselor from California.

The treatment was conducted in Watkins's home and videotaped. According to trial transcripts, Watkins and Ponder conducted more than four days of "holding therapies," where they grabbed or covered Candace's face 138 times, shook or bounced her head 392 times, and shouted into her face 133 times. When this failed to break her, they put tiny 68-pound Candace inside a flannel sheet and covered her with sofa pillows, while several adults (with a combined weight of nearly 700 pounds) lay on top of her so she could be "reborn." Ponder told Candace that she was "a teeny little baby" in the womb, commanding her to "come out head first" and "push with your feet." In response, Candace screamed, "I can't breathe, I can't do it! Somebody's on top of me. I want to die now! Please! Air!"

According to AT theory, Candace's reaction was a sign of her emotional resistance; she needed more confrontation to reach the rage necessary to break through the wall and achieve emotional healing. ACE (now operating as the Institute for Attachment and Child Development), for example, claims that "confrontation is sometimes necessary to break through a child's defenses and reach the hurting child within. Confrontation of faulty thinking patterns and destructive behavior patterns is essential if change is to occur."

Putting theory into practice, Ponder admonished, "You're gonna die." Candace begged "Please, please, I can't breathe." Ponder instructed the others to "press more on top," on the premise that AD children exaggerate their distress. Candace vomited, then cried "I gotta poop." Her mother entreated, "I know it's hard but I'm waiting for you."

After forty minutes Candace went silent. Ponder rebuked her "Quitter, quitter!" Someone joked about performing a C-section, while Ponder patted a dog that meandered by. After thirty minutes of silence, Watkins sarcastically remarked, "Let's look at this twerp and see what's going on—is there a kid in there somewhere? There you are lying in your own vomit—aren't you tired?"

Candace wasn't tired; she was dead. "This ten-year-old child died of cerebral edema and herniation caused by hypoxic-ischemic encephalopathy," the autopsy report clinically stated. The proximate cause of Candace's death was suffocation, and each of her therapists received the minimum sentence of sixteen years for "reckless child abuse resulting in death." The ultimate cause was pseudoscientific quackery masquerading as psychological science. “However bizarre or idiosyncratic these treatments appear—and however ineffective or harmful they may be to children—they emerge from a complex internal logic, based, unfortunately, on faulty premises,” write Jean Mercer, Larry Sarner, and Linda Rosa in their penetrating 2003 analysis *Attachment Therapy on Trial* (Praeger; see also www.ChildrenInTherapy.org). These therapists killed Candace not because they were evil but because they were in the grip of a pseudoscientific theory.

AT continues to flourish. ATTACh claims to have about six hundred members. The numbers may be even higher, the authors say, because the practice goes by so many different labels, including “Holding-Nurturing Process, holding therapy, rage reduction, cuddle time, gentle containment, compression therapy, safe containment,” and others.

By whatever name, AT is a pseudoscience that kills and tortures children. It should be condemned as unethical and banned as illegal before it kills and tortures again.

26

Cures and Cons

Natural scams "he" doesn't want you to know about

One route to social commentary is over-the-top comedic parody, a technique effectively employed in the fake commercials on NBC's *Saturday Night Live*. A memorable example that aired during the 1980s' craze for high-fiber diets was "Colon Blow cereal," equivalent to thirty thousand bowls of oat bran cereal. In the same genre was Dan Aykroyd's slimy pitchman character Irvin Mainway, who hawked items on late-night television such as the "Bass-O-Matic blender" and toys like "Bag O' Glass" and a Johnny Switchblade Punk action figure.

Imagine Irvin Mainway pitching Colon Blow cereal blended with Bass and Glass, and you've got a fair picture of Kevin Trudeau who, up to 139 times in one week, pitches late-night viewers about his self-published book *Natural Cures "They" Don't Want You to Know About*, a rambling farrago of uninformed opinions, conspiracy theories, and cheeky jabs at medical, pharmaceutical, and governmental authorities ("they"). The book is so risibly ridiculous that surely no people, not even the most desperately ill, would take it seriously, would they?

Apparently they would, to the tune of millions of copies sold, elevating the book to the *New York Times*'s best-seller list. If readers had purchased Trudeau's Mega Memory System perhaps they would have remembered that the perfidious huckster spent two years in federal prison after pleading guilty to credit card fraud, and that the Federal Trade Commission (FTC) banned Trudeau "from appearing in, producing, or disseminating future infomercials that advertise any type of product, service, or program to the public, except for truthful infomercials for informational publications. In addition, Trudeau cannot make disease or health benefits claims for any type of product, service, or program in any advertising, including print, radio, Internet, television, and direct mail solicitations, regardless of the format and duration." Trudeau had to pay $500,000 in consumer redress for his bogus infomercials, and another $2 million to settle charges against him for claiming that coral calcium cures cancer (it doesn't) and that an analgesic product called Biotape permanently relieves pain (it doesn't).

Amazingly, *Natural Cures* is exempt from this injunction. "Books are fully protected speech. He can author a book and voice his opinions," says Heather Hippsley, assistant director for the division of advertising practices at the FTC who investigated Trudeau's infomercials "The line is: Informational materials, OK. Products and services, banned."

So Trudeau is free to dole out in print such opinions as these: "Medical science has absolutely, 100 percent, failed in the curing and prevention of illness, sickness, and disease." (Smallpox is not a disease?) "Get all metal out of your dental work." (Won't this *help* the medical cartel?) "Sun block has been shown to cause cancer." (References?) "Don't drink tap water." (Wrong: studies show it is as safe as bottled water.) "Animals in the wild never get sick." (No need to worry about avian influenza.) "Get 15 colonics in 30 days." (Can I bring a friend?) "Wear white. The closer you get to white, the more positive energy you bring into your energetic field." (Why is Trudeau wearing all black on the book cover?) "Stop taking nonprescription and prescription drugs." (Including insulin for diabetes?) "This includes vaccines." (Welcome back, polio.) "Have sex." (Without prescription Viagra?)

This six-hundred-page medical advice book contains no index, no bibliography, and no references. In their stead are testimonials and ads for the audio edition, the newsletter, and a sequel in the works about "weight loss secrets 'they' don't want you to know about."

As for the "natural cures" themselves, some are not cures at all but just obvious healthy lifestyle suggestions: eat less, exercise more, reduce stress. Some of the natural cures are flat out wrong, such as oral chelation for heart disease, while others are laughably ludicrous, such as a magnetic mattress pad and crocodile protein peptide for fibromyalgia. Worst of all are all the natural cures that the book directs the reader to Trudeau's web page to find. When you go there, however, and click on a disease, to get the cure you first have to become a web site member at $499 lifetime or $9.95 a month. It's the con man's combo: bait and switch (the book directs them to the web page) and double dipping (sell them the book, then sell them the membership).

Why don't "they" want you to know about these natural cures? "Money and Power," says Trudeau. "Most people have no idea just how powerful a motivating force money and power can be." Kevin Trudeau certainly does, and this book is a testimony to that fact.

There is one medical lesson that I gleaned from this otherwise feckless author, well expressed in an old Japanese proverb: "*Baka ni tsukuru kusuri wa nai desu*"—"There is no medicine that cures stupidity." *Domo arigato*, Mr. Trudeau.

Update: On March 17, 2014, Kevin Trudeau was sentenced to ten years in prison on a number of charges, including criminal contempt for lying in his infomercials about his book The Weight Loss Cure "They" Don't Want You to Know About, *and for failing to pay anything toward a $37.6 million fine imposed by the Federal Trade Commission.*

IV

THE PARANORMAL AND THE SUPERNATURAL

27

Deconstructing the Dead

"Crossing over" to expose the tricks of popular spirit mediums

Humans are pattern-seeking, storytelling animals. Like all other animals, we evolved to connect the dots between events in nature to discern meaningful patterns for our survival. Like no other animals, we tell stories about the patterns we find. Sometimes the patterns are real, sometimes they are illusions.

An illusion of a meaningful pattern based on anecdotes that has generated countless stories is the alleged ability of mediums to talk to the dead. The hottest medium today is former ballroom-dance instructor John Edward, star of the hit Sci Fi Network television series *Crossing Over* and author of the *New York Times* best-selling book *One Last Time*. His show is so hot that he is about to go opposite Oprah in CBS's afternoon lineup.

How does Edward appear to talk to the dead? In short, it's a trick. Edward begins by selecting a section of the studio audience of about twenty people, saying something like "I'm getting a George over here. I don't know what this means. George could be someone who passed over, he could be someone here, he could be someone that you know," etc. Of course, such generalizations lead to a "hit." Now that he's targeted his mark (the street con's term for the person about to be taken), the "reading" begins, utilizing three techniques:

1. Cold reading, where you literally "read" someone "cold," knowing nothing about them. You ask lots of questions and make numerous statements and see what sticks. "I'm getting a P name. Who is this, please?" "He's showing me something red. What is this, please?" And so on. Most statements are wrong. If subjects have time they visibly shake their heads "no." But Edward is so fast that they usually only have time to acknowledge the hits. And as B. F. Skinner showed in his experiments on superstitious behavior, subjects only need an occasional reinforcement to be convinced there is a real pattern (slot machines need only pay off infrequently to keep people involved). In an exposé I did for WABC New York, we counted about one statement per second in the opening minute as he riffled through names, dates, colors, diseases, conditions, situations, relatives, and the like. It goes so fast that you have to stop the tape and go back to catch them all.

2. Warm reading, utilizing known principles of psychology that apply to nearly everyone. Many grieving people wear a piece of jewelry that has a connection to their loved one. Following the death of her husband, for example, Katie Couric on *The Today Show* wore his ring on a necklace. Mediums know this about mourning people and will say something like "do you have a ring or a piece of jewelry on you, please?" Edward is also facile at determining the cause of death by focusing either on the chest or head areas, and then exploring whether it was a slow or a sudden end. He works his way through the half dozen major causes of death in rapid-fire manner. "He's telling me there was a pain in the chest." If he gets a positive nod, he continues. "Did he have cancer, please? Because I'm seeing a slow death here." If he gets the nod, he takes the hit. If the subject hesitates, he will quickly shift to heart attack. If it is the head, he goes for stroke or head injury from an automobile accident or fall.

3. Hot reading, where the medium obtains information on a subject ahead of time. One man who got a reading on Edward's show reports that "his 'production assistants' were always around while we waited to get into the studio. They told us to keep very quiet, and they overheard a lot. I think that the whole place is bugged somehow. Also, once in the studio we had to wait around for almost two hours before the show began. Throughout that time everybody was talking about what dead relative of theirs might pop up. Remember that all this occurred under microphones and with cameras already set up."

Most of the time, however, mediums do not need to cheat. They are successful because they are dealing with a subject the likes of which would be hard to top for tragedy and finality—death. Sooner or later we all will face this inevitability, and we are often at our most vulnerable at such times. Giving deep thought to this reality can cause the most controlled and rational among us to succumb to our emotions.

This is the reason why mediums are unethical and dangerous. They are preying on the emotions of grieving people, and, as grief counselors know, the best way to deal with death is to face it head on. Death is a part of life, and pretending that the dead are gathering in a television studio in New York to talk twaddle with a former ballroom-dance instructor is an insult to the intelligence and humanity of the living.

28

Psychic Drift

Why most scientists do not believe in ESP and psi phenomena

Throughout the first half of the nineteenth century the theory of evolution remained mired in conjecture until Charles Darwin and Alfred Russel Wallace compiled a massive body of evidence and discovered a mechanism—natural selection—to explain what would power the evolutionary machine.

Throughout the first half of the twentieth century the theory of continental drift—proposed in 1915 by the German scientist Alfred Wegener—hovered on the margins of science until the 1960s and the discovery of midoceanic ridges; geomagnetic patterns corresponding to continental plate movement; and, most importantly, plate tectonics as the motor that drives continents.

Data and theory. Evidence and mechanism. These are the twin pillars of sound science. Without data and evidence there is nothing for a theory or mechanism to explain. Without a theory and mechanism, data and evidence drift aimlessly on a boundless sea.

For more than a century now claims have been made for the existence of psi, or psychic phenomena. In the late nineteenth century, organizations like the Society for Psychical Research were founded to employ rigorous scientific methods in the study of psi, and they had many world-class scientists in support, including none other than Wallace (Darwin was skeptical). In the twentieth century, psi periodically found its way into serious academic research programs, from Joseph Rhine's Duke University experiments in the 1920s to Daryl Bem's Cornell University research in the 1990s.

In January 1994, for example, Bem and his late University of Edinburgh parapsychologist colleague Charles Honorton published "Does Psi Exist? Replicable Evidence for an Anomalous Process of Information Transfer" in the prestigious review journal *Psychological Bulletin*. Conducting a meta-analysis of forty published experiments, the authors concluded, "the replication rates and effect sizes achieved by one particular experimental method, the ganzfeld procedure, are now sufficient to warrant bringing this body of data to the attention of the wider psychological community." (A meta-analysis is a statistical technique that combines the results from many studies to look for an overall effect, even if the results from the individual studies were insignificant; the ganzfeld procedure places the "receiver" in a sensory isolation room with Ping-Pong ball halves over the eyes, headphones playing white noise over the ears, and the "sender" in another room psychically transmitting photographic or video images.)

Despite finding evidence for psi (subjects had a hit rate of 35 percent when 25 percent was expected by chance), Bem and Honorton lamented, "Most academic psychologists do not yet accept the existence of psi, anomalous processes of information or energy transfer (such as telepathy or other forms of extrasensory perception) that are currently unexplained in terms of known physical or biological mechanisms."

Why don't scientists accept psi? Daryl Bem has a stellar reputation as a rigorous experimentalist and he has presented us with statistically significant results. Aren't scientists supposed to be open to changing their minds when presented with new data and evidence? The reason for skepticism is that we need both replicable data and a viable theory, both of which are missing in psi research.

Data. Both the meta-analysis and ganzfeld techniques have been challenged. Ray Hyman from the University of Oregon found inconsistencies in the experimental procedures used in different ganzfeld experiments (that were lumped together in Bem's meta-analysis as if they used the same procedures), and that the statistical test employed (Stouffer's *Z*) was inappropriate for such a diverse data set. He also found flaws in the target randomization process (the sequence the visual targets were sent to the receiver), resulting in a target selection bias: "All of the significant hitting was done on the second or later appearance of a target. If we examined the guesses against just the first occurrences of targets, the result is consistent with chance." Richard Wiseman from the University of Hertfordshire conducted a meta-analysis of thirty more ganzfeld experiments and found no evidence for psi, concluding that psi data are nonreplicable. Bem countered with ten additional ganzfeld experiments he claims are significant, and he has additional research he plans to publish. And so it goes . . . with more to come in the data debate.

Theory. The deeper reason why scientists remain skeptical of psi—and will remain so even if more significant data are published—is that there is no explanatory theory for how psi works. Until psi proponents can explain how thoughts generated by neurons in the sender's brain can pass through the skull and into the brain of the receiver, skepticism is the appropriate response, as it was for evolution sans natural selection, and continental drift without plate tectonics. If the data show that there is such a phenomenon as psi that needs explaining (and I am not convinced that it does), then we still need a causal mechanism.

Until psi finds its Darwin it will continue to drift on the borderlands of science.

29

Demon-Haunted Brain

If the brain mediates all experience, then paranormal phenomena are nothing more than neuronal events

Five centuries ago demons haunted our world, with incubi and succubi tormenting their victims as they lay asleep in their beds. Two centuries ago spirits haunted our world, with ghosts and ghouls harassing their sufferers at all hours of the night. In the twentieth century aliens haunted our world, with grays and greens abducting captives out of their beds and whisking them away for probing and prodding. Today people are experiencing out-of-body experiences, floating above their beds, out of their bedrooms, and even off the planet into space.

What is going on here? Are these elusive creatures and mysterious phenomena in our world or in our minds? New evidence indicates that they are, in fact, products of the brain. Neuroscientist Michael Persinger, in his laboratory at Laurentian University in Sudbury, Canada, for example, can induce all of these experiences in subjects by subjecting their temporal lobes to patterns of magnetic fields (I tried it and had a mild out-of-body experience).

Similarly, the September 19, 2002, issue of *Nature* reported that the Swiss neuroscientist Olaf Blanke and his colleagues discovered that they could bring about out-of-body experiences through electrical stimulation of the right angular gyrus in the temporal lobe of a forty-three-year-old woman suffering from severe epileptic seizures. In initial mild stimulations she reported "sinking into the bed" or "falling from a height." More intense stimulation led her to "see myself lying in bed, from above, but I only see my legs and lower trunk." Another stimulation induced "an instantaneous feeling of 'lightness' and 'floating' about two meters above the bed, close to the ceiling."

In a related study reported in the 2001 book *Why God Won't Go Away*, researchers Andrew Newberg and Eugene D'Aquili found that when Buddhist monks meditate and Franciscan nuns pray, their brain scans indicate strikingly low activity in the posterior superior parietal lobe, a region of the brain the authors have dubbed the Orientation Association Area (OAA), whose job it is to orient the body in physical space (people with damage to this area have a difficult time negotiating their way around a house). When the OAA is booted up and running smoothly there is a sharp distinction between self and nonself. When the OAA is in sleep mode—as in deep meditation and prayer—that division breaks down, leading to a blurring of the lines between reality and fantasy, between feeling in body and out of body. Perhaps this is what happens to monks who experience a sense of oneness with the universe, or with nuns who feel the presence of God, or with alien abductees floating out of their beds up to the mother ship.

Sometimes trauma can trigger such experiences. The December 2001 issue of *Lancet* published a Dutch study in which of 344 cardiac patients resuscitated from clinical death, 12 percent reported near-death experiences, where they had an out-of-body experience and saw a light at the end of a tunnel. Some even described speaking to dead relatives. Since our normal experience is of stimuli coming into the brain from the outside, when a part of the brain abnormally generates these illusions another part of the brain interprets them as external events. Hence the abnormal is thought to be the paranormal.

These studies are only the latest to deliver blows against the belief that mind and spirit are separate from brain and body. In reality, all experience is mediated by the brain. Large brain areas such as the cortex coordinate inputs from smaller brain areas such as the temporal lobes, which themselves collate neural events from still smaller brain modules, like the angular gyrus. This reduction continues all the way down to the single-neuron level, where highly selective neurons, sometimes described as "grandmother" neurons, fire only when subjects see someone they know. Caltech neuroscientists Christof Koch and Gabriel Kreiman, in conjunction with UCLA neurosurgeon Itzhak Fried, have even found a single neuron that fires when the subject is shown a photograph of Bill Clinton. The Monica neuron must be closely connected.

Of course, we are not aware of the workings of our own electrochemical systems. What we actually experience is what philosophers call qualia, or subjective states of thoughts and feelings that arise from a concatenation of neural events.

It is the fate of the paranormal and the supernatural to be subsumed into the normal and the natural. In fact, there is no paranormal or supernatural; there is only the normal and the natural . . . and mysteries yet to be explained. It is the job of science, not pseudoscience, to solve those puzzles with natural, not supernatural, explanations.

30

Codified Claptrap

The Bible code is numerological nonsense masquerading as science

In the epilogue of *In Memoriam A.H.H.*, Alfred, Lord Tennyson captured the essence of the quest for a single unifying principle and purpose in nature:

> *One God, one law, one element,*
> *And one far-off divine event,*
> *To which the whole creation moves.*

From Hal Lindsey's 1970s blockbuster *The Late Great Planet Earth* (more than thirty-five million sold) to today's *Left Behind* series by Tim LaHaye and Jerry Jenkins (more than fifty million sold), the noble dream of uniting microcosm and macrocosm and finding teleological succor in the march of time has become big business. And since we live in the Age of Science, if you can sprinkle your homiletics with scientistic jargon, so much the better. The latest and most egregious example of the misuse of science in the (dis)service of religion can be found in Michael Drosnin's *Bible Code II: The Countdown*, enjoying a lucrative ride on the *New York Times* best-seller list, as did the 1997 original.

According to proponents of the Bible code (poseurs, in my view, of a long and honorable tradition of biblical exegesis), itself a subset of the genre of biblical numerology and Kabalistic mysticism popular since the Middle Ages, the Hebrew Pentateuch (first five books of the Bible) can be decoded through an equidistant letter sequencing computer program by taking every *n*th letter, where *n* equals whatever number you wish: 7, 19, 3,027. . . . Print out that string of letters in a block of type, then search left to right, right to left, up to down, down to up, and diagonally in any direction for any interesting patterns. Seek and ye shall find.

Predictably, in 1997 Drosnin "discovered" such current events as Yitzhak Rabin's assassination; Netanyahu's election; comet Shoemaker-Levy's collision with Jupiter; Timothy McVeigh and the Oklahoma City bombing; and, of course, the end of the world in 2000. Since the world did not end and current events dated his first book, Drosnin continued the search and found—lo and behold—that the Bible predicted the Bill and Monica tryst; the Bush/Gore election debacle; and, of course, the World Trade Center cataclysm.

Just like the prophecies of soothsayers past and present, all such predictions are actually postdictions (note that not one psychic or astrologer forewarned us about 9/11). To be tested scientifically Bible coders would need to predict events *before* they happen. They won't, because they can't—as Danish physicist Niels Bohr averred, predictions are difficult, especially about the future. Instead, in 1997 Drosnin proposed this test of his thesis: "When my critics find a message about the assassination of a prime minister encrypted in *Moby Dick*, I'll believe them."

Australian mathematician Brendan McKay did just that, finding no less than thirteen political assassinations secreted in the great novel, along with additional finds in *War and Peace* and other tomes (http://bit.ly/RwaXcX). Humorously, American mathematician David Thomas predicted the Chicago Bulls' NBA championship in 1998 from his code search of Tolstoy's novel, and recently found "The Bible Code is a silly, dumb, fake, false, evil, nasty, dismal fraud and snake-oil hoax" concealed in the first chapter of *Bible Code II* (http://bit.ly/1wUu3Mx)!

If there is an encrypted message in all this numerological poppycock it is this: there is a deep connection between how the mind works and how the world works. We are pattern-seeking animals, descendants of those hominids who were especially dexterous at making causal connections between events in nature. The associations were real often enough that the process became ingrained in our neural architecture. Unfortunately, the belief engine sputters occasionally, identifying false patterns as real. The process may not be enough to prevent you from passing on your genes for finding false positives into the next generation, but it does create superstitious and magical thinking. This process is coupled to the law of large numbers that accompanies our complex world where, as it is said, million-to-one odds happen eight times a day in New York City.

Given our propensity to find patterns in a superfluity of data, is it any wonder that so many are taken in by such codified claptrap? The problem is present, pervasive, and a permanent part of our cognitive machinery. The solution is science—our preeminent pattern-discriminating method and our best hope for detecting a real signal within the noise of nature's cacophony.

Update: Not surprisingly, there is now a Bible Code III, *updated to include predictions about President Barack Obama and others:* http://bit.ly/1nLO66f

31

The Myth Is the Message

Yet another discovery of the lost continent of Atlantis shows why science and myth make uneasy bedfellows

Myths are stories that express meaning, morality, or motivation. Whether they are true or not is irrelevant. But because we live in the Age of Science, we have a preoccupation with corroborating our myths.

Consider biblical archaeology, which goes in search of data to confirm or refute the Bible's mythic tales. Some appear to have a basis in fact (e.g., King David); others have not a shred of extrabiblical evidence (e.g., Moses). But what can this marriage of science and myth really accomplish? Some have speculated, for example, that the volcanic eruption that destroyed the island of Santorini in the Aegean Sea about thirty-five hundred years ago may account for the biblical plagues that befell the Egyptians and aided Moses. But if a geological paroxysm explains biblical miracles, doesn't that turn God into a divine volcanologist?

This same event has been proffered as the convulsion that sank the continent of Atlantis. Unless the Bronze Age Minoan civilization that flourished there counts as the vanished Atlanteans, however, believers will have to search elsewhere. They have. On June 6, 2004, the BBC released a story about satellite images allegedly locating Atlantis in the South of Spain (http://bbc.in/1tOvhVW). According to Rainer Kuehne, from the University of Wuppertal in Germany, "Plato wrote of an island of five stades (925m) diameter that was surrounded by several circular structures—concentric rings—some consisting of Earth and the others of water. We have in the photos concentric rings just as Plato described."

Kuehne reported his findings in the online edition of the journal *Antiquity*, claiming to have identified two rectangular structures surrounded by concentric rings near the city of Cádiz. Kuehne suggests that the structures match the description in Plato's *Critias* of the silver and golden temples devoted to Cleito and Poseidon, and that the high mountains of Atlantis are actually those of the Sierra Morena and Sierra Nevada. "Plato also wrote that Atlantis is rich in copper and other metals," he added. "Copper is found in abundance in the mines of the Sierra Morena."

Atlantis has been "found" in numerous local areas, including the Mediterranean, the Canaries, the Azores, the Caribbean, Tunisia, West Africa, Sweden, Iceland, and even South America. But what if there is nothing to find? What if Plato made up the story for mythic purposes? He did. Atlantis is a tale about what happens to a civilization when it becomes combative and corrupt. Plato's purpose was to warn his fellow Athenians to pull back from the precipice that war and wealth were taking them over.

In the *Timaeus*, Plato's dialogist, Critias, explains that Egyptian priests told the Greek wise man Solon that his ancestors once defeated a mighty empire just beyond the "Pillars of Hercules" (usually identified by Atlantologists as the Straits of Gibraltar), after which "there were violent earthquakes and in a single day and night all sank into the earth and the Island of Atlantis in like manner disappeared into the depths of the sea." Critias describes the city as a series of circular canals lined with colorful palaces adorned in gold. Poseidon resided in a silver temple with an ivory roof, and a racecourse was built between the canals. Atlantean wealth afforded a military-industrial complex of 10,000 chariots, 24,000 ships, 60,000 officers, 120,000 hoplites, 240,000 cavalry, and 600,000 archers and javelin throwers (your myth detection alarm should be going off about now). Corrupted by excessive belligerence and avarice, Zeus called forth the other gods to his home, "and when he had gathered them there he said . . ." The sentence ends there. Plato had made his point.

The fodder for Plato's imagination came from his experiences growing up at the terminus of Greece's Golden Age, brought about, in part, by the costly wars against the Spartans and the Carthaginians. He visited cities such as Syracuse, featuring numerous Atlantean-like temples, and Carthage, whose circular harbor was controlled from a central island. Earthquakes were common: when he was fifty-five one leveled the city of Helice, only forty miles from Athens, and, most tellingly, the year before he was born an earthquake flattened a military outpost on the small island of Atalantë.

Plato wove historical fact into literary myth: "We may liken the false to the true for the purpose of moral instruction." The myth is the message.

32

Turn Me On, Dead Man

What do the Beatles, the Virgin Mary, Jesus, Patricia Arquette, and Michael Keaton all have in common?

In September 1969, as I began ninth grade, a rumor circulated that the Beatles' Paul McCartney was dead, killed in a 1966 automobile accident and replaced by a look-alike. The clues were there in the albums, if you knew where to look. Seek and ye shall find.

Sgt. Pepper's "A Day in the Life," for example, recounts the accident: *He blew his mind out in a car / He didn't notice that the lights had changed / A crowd of people stood and stared / They'd seen his face before / Nobody was really sure if he was from the House of Lords [or "Paul," according to some].* The cover of the *Abbey Road* album, released late that September, shows the Fab Four walking across a street in what looks like a funeral procession, with John in white as the preacher, Ringo in black as the pallbearer, a barefoot and out-of-step Paul as the corpse, and George in work clothes as the grave digger. In the background is a Volkswagen Beetle (!) whose license plate reads 28IF—Paul's age "if" he had not died in the 1966 accident.

Spookiest of all were the clues embedded in songs played backward. On a cheap turntable I moved the speed switch midway between 33 1/3 and 45 to disengage the motor drive, then manually turned the record backward and listened in wide-eared wonder. The eeriest is "Revolution

No. 9" from the *White Album*, in which an ominously deep voice endlessly repeats *number nine . . . number nine . . . number nine . . . .* Played backward you hear *turn me on, dead man . . . turn me on, dead man . . . turn me on, dead man. . . .*

In time, thousands of clues emerged as the rumor mill cranked up (Google "Paul is dead" for examples), despite John Lennon's 1970 statement to *Rolling Stone* magazine: "That was all bullshit, the whole thing was made up." But made up by whom? Not the Beatles, so this was the art of public pattern-seeking and media conspiracy-mongering at its finest.

What we have here is a signal/noise problem. Humans evolved brains that are pattern-recognition machines, designed to detect signals that enhance or threaten survival amid a very noisy world. Also known as association learning (associating A and B as causally connected), we are very good at it, or at least good enough to have survived and passed on the genes for the capacity of association learning. Unfortunately, the system has flaws. Superstitions are false associations—A appears to be connected to B, but it is not (the baseball player who doesn't shave and hits a home run). Las Vegas was built on false association learning.

Consider a few recent examples of false pattern-recognition (Google key words for visuals): the face of the Virgin Mary on a grilled-cheese sandwich (looks more like Greta Garbo to me); the face of Jesus on an oyster shell (resembles Charles Manson, I think); the hit NBC television series *Medium*, in which Patricia Arquette plays psychic Allison Dubois, whose occasional thoughts and dreams seem connected to real-world crimes; the popular film *White Noise*, in which Michael Keaton's character believes he is receiving messages from his dead wife through tape recorders and other electronic devices in what is called EVP, or Electronic Voice Phenomenon. EVP is another version of what I call TMODMP, the Turn Me On, Dead Man Phenomenon—if you scan enough noise you will eventually find a signal, whether it is there or not.

Anecdotes fuel pattern-seeking thought. Aunt Mildred's cancer went into remission after she imbibed extract of seaweed—maybe it works. But there is only one surefire method of proper pattern-recognition, and that is science. Only when a group of cancer patients taking seaweed extract is compared to a control group can we draw a valid conclusion.

We evolved as a social primate species whose language ability facilitated the exchange of such association anecdotes. The problem is that although true pattern-recognition helps us survive, false pattern-recognition does not necessarily get us killed, and so the overall phenomenon endured the winnowing process of natural selection. The Darwin Awards (honoring those who remove themselves from the gene pool "in really stupid ways"), like this essay, will never want for examples. Anecdotal thinking comes naturally; science requires training.

33

Rupert's Resonance

The theory of "morphic resonance" posits that people have a sense of when they are being stared at. What does the research show?

Have you ever noticed how much easier it is to do a newspaper crossword puzzle later in the day? Me neither. But according to Rupert Sheldrake it is because the collective wisdom of the morning successes resonates throughout the cultural morphic field.

The theories of Rupert Sheldrake, a Cambridge University–trained scientist and scion of paranormal ancestors dating back to Alfred Russel Wallace, proffer scientistic unity through what he calls morphic resonance—similar forms (morphs) resonate and exchange information through a universal life force—"the basis of memory in nature . . . the idea of mysterious telepathy-type interconnections between organisms and of collective memories within species." As Rupert recounted in a 1999 Salon.com interview: "Descartes believed the only kind of mind was the conscious mind. Then Freud reinvented the unconscious. Then Jung said it's not just a personal unconscious but a collective unconscious. Morphic resonance shows us that our very souls are connected with those of others and bound up with the world around us." Descartes, Freud, Jung, and Rupert. Resonance indeed.

As well, Sheldrake's morphic resonance also means that "as time goes on, each type of organism forms a special kind of cumulative collective memory," he writes in his 1981 book *A New Science of Life* (J. P. Tarcher). "The regularities of nature are therefore habitual. Things are as they are because they were as they were." In this book and subsequent ones—*The Presence of the Past,* HarperCollins, 1988; *The Physics of Angels*, with Matthew Fox, HarperSanFrancisco, 1996—Sheldrake, a trained botanist and onetime research fellow of the Royal Society, details the theory, which was hotly debated in the June 2005 issue of the *Journal of Consciousness Studies.*

Morphic resonance, says Sheldrake, is "the idea of mysterious telepathy-type interconnections between organisms and of collective memories within species" and explains phantom limbs, homing pigeons, how dogs know when their owners are coming home, and such psychic phenomena as how people know when someone is staring at them. "Vision may involve a two-way process, an inward movement of light and an outward projection of mental images," Sheldrake explains. Thousands of trials conducted by anyone who downloaded the experimental protocol from Sheldrake's web page "have given positive, repeatable, and highly significant results, implying that there is indeed a widespread sensitivity to being stared at from behind."

Let's examine this claim more closely. First, science is not normally conducted by strangers who happen upon a web page protocol, so we have no way of knowing if these amateurs controlled for intervening variables and experimenter biases. Second, psychologists dismiss anecdotal accounts of this sense to a reverse self-fulfilling effect: a person suspects being stared at and turns to check; such head movement catches the eyes of would-be starers, who then turn to look at the staree, who thereby confirms the feeling of being stared at.

Third, in 2000 John Colwell from Middlesex University, London, conducted a formal test utilizing Sheldrake's suggested experimental protocol, with twelve volunteers who participated in twelve sequences of twenty stare or no-stare trials each, with accuracy feedback provided for the final nine sessions. Results: subjects were able to detect being stared at only when accuracy feedback was provided, which Colwell attributed to the subjects learning what was, in fact, a nonrandom presentation of the experimental trials. When the University of Hertfordshire psychologist

Richard Wiseman also attempted to replicate Sheldrake's research, he found that subjects detected stares at rates no better than chance.

Fourth, the confirmation bias (where we look for and find confirmatory evidence for what we already believe) may be at work here. In the special issue of the *Journal of Consciousness Studies* devoted to "Sheldrake and His Critics," I rated the fourteen open peer commentaries on Sheldrake's target article (on the sense of being stared at) on a scale of 1 to 5 (critical, mildly critical, neutral, mildly supportive, supportive). Without exception, the 1s, 2s, and 3s were all traditional scientists from mainstream institutions, whereas the 4s and 5s were all affiliated with fringe and proparanormal institutions.

Fifth, there is an experimenter bias problem. Institute of Noetic Sciences researcher Marilyn Schlitz (a believer in psi) collaborated with Wiseman (a skeptic of psi) in replicating Sheldrake's research and discovered that when *they* did the staring Schlitz found statistically significant results, whereas Wiseman found chance results.

Sheldrake responds that skeptics dampen the morphic field's subtle power, whereas believers enhance it. Of Wiseman, Sheldrake remarked: "Perhaps his negative expectations consciously or unconsciously influenced the way he looked at the subjects."

Perhaps, but wouldn't that mean that this claim is ultimately nonfalsifiable? If both positive and negative results are interpreted as supporting a theory, how can we test its validity? Skepticism is the default position because the burden of proof is on the believer, not the skeptic.

34

Mr. Skeptic Goes to Esalen

Science and spirituality on the California coast

The Esalen Institute is a cluster of meeting rooms, lodging facilities, and hot tubs all nestled into a stunning craggy coastal outcrop of the Pacific Ocean in Big Sur, California. In his 1985 book *Surely You're Joking, Mr. Feynman*, the Nobel laureate physicist Richard Feynman recounted his experience in the natural hot spring baths there, in which a woman was being massaged by a man she just met. "He starts to rub her big toe. 'I think I feel it,' he says. 'I feel a kind of dent—is that the pituitary?' I blurt out, 'You're a helluva long way from the pituitary, man!' They looked at me horrified and said, 'It's reflexology!' I quickly closed my eyes and appeared to be meditating."

With that as my introduction to the mecca of the New Age movement, I accepted an invitation to host a weekend workshop there on science and spirituality. Given my propensity for skepticism when it comes to most of the paranormal piffle proffered by the *prajna* peddlers meditating and soaking their way to nirvana here, I was surprised the hall was full. Perhaps skeptical consciousness is rising!

It was in the extracurricular conversations, however, during healthy homegrown meals and while soaking in the hot tubs, that I gleaned a sense of what people believe and why. Once it became known that Mr. Skeptic was there, for example, I heard one after another "how do you explain *this*?" story, mostly involving angels, aliens, and the usual paranormal fare. But this being Esalen—ground zero for all that is weird and wonderful in the human potential movement—there were some singularly unique accounts.

One woman explained the theory behind "bodywork," a combination of massage and "energy work" that involves adjusting the body's seven energy centers called chakras. I signed up for a massage, which was the best I've ever had (and I had a lot when I was a bike racer), but when another practitioner told me about how she cured a woman's migraine headache by directing a light beam through her head, I decided that practice and theory are best kept separate. Another woman warned about the epidemic of Satanic cults. "But there's no evidence of such cults," I countered. "Of course not," she explained. "They erase all memories and evidence of their nefarious activities."

One gentleman recounted a lengthy tantric sexual encounter with his lover that lasted for many hours, at the culmination of which a lightning bolt shot through her left eye followed by a being of blue light, a child entering her womb, ensuring conception. Nine months later, friends and gurus joined the couple in a hothouse, sweating their way through their own "rebirthing" process before the mother gave birth to a baby boy. The father then told him that he would need to become an athlete to get into college; two decades later, this young man became a professional baseball player. "How do you explain *that*?" I was asked. I quickly closed my eyes and appeared to be meditating.

People have and share such experiences, and impart larger significance to them, because we have a cortex large enough to conceive of such transcendent notions, and an imagination creative enough to concoct fantastic narratives. If we define the spirit (or soul) as the pattern of information of which we are made—our genes, proteins, memories, and personalities—then spirituality is the quest to know the place of our essence within the deep time of evolution and the deep space of the cosmos.

There are many ways to be spiritual, and science is one in its awe-inspiring account about who we are and where we came from. "The cosmos is within us. We are made of star stuff. We are a way for the cosmos to know itself," began the late astronomer Carl Sagan in the opening scene of *Cosmos*, filmed just down the coast from Esalen, in referring to the stellar origins of the chemical elements of life. "We've begun at last to wonder about our origins, star stuff contemplating the stars, organized collections of ten billion billion billion atoms contemplating the evolution of matter, tracing that long path by which it arrived at consciousness here on the planet Earth and perhaps throughout the cosmos. Our obligation to survive and flourish is owed not just to ourselves but also to that cosmos, ancient and vast, from which we spring."

That is spiritual gold.

V

ALIENS AND UFOS

35

Shermer's Last Law

Any sufficiently advanced extraterrestrial intelligence is indistinguishable from God

As scientist extraordinaire (most profoundly as inventor of the communications satellite) and author of an empire of science fiction books and films (most notably *2001: A Space Odyssey*), Arthur C. Clarke is one of the most far-seeing visionaries of our time. Thus his pithy quotations tug harder on our collective psyches for their inferred insights into humanity and our place in the cosmos. And none do so more than his famous three laws:

Clarke's First Law: "When a distinguished but elderly scientist states that something is possible he is almost certainly right. When he states that something is impossible, he is very probably wrong."

Clarke's Second Law: "The only way of discovering the limits of the possible is to venture a little way past them into the impossible."

Clarke's Third Law: "Any sufficiently advanced technology is indistinguishable from magic."

This last observation stimulated me to think more on the relationship of science and religion, particularly the impact the discovery of an extraterrestrial intelligence (ETI) would have on both traditions. To that end I would like to immodestly propose Shermer's Last Law (I don't believe in naming laws after oneself, so as the Good Book warns, the last shall be first and the first shall be last): *Any sufficiently advanced ETI is indistinguishable from God.*

God is typically described by Western religions as omniscient and omnipotent. Since we are far from the mark on these traits, how could we possibly distinguish a God who has them absolutely, from an ETI who has them in relatively (to us) copious amounts? Thus we would be unable to distinguish between absolute and relative omniscience and omnipotence. But if God were only relatively more knowing and powerful than we, then by definition it *would* be an ETI! Consider two observations and one deduction:

1. Biological evolution operates at a snail's pace compared to technological evolution (the former is Darwinian and requires generations of differential reproductive success; the latter is Lamarckian and can be implemented within a single generation).

2. The cosmos is very big and space is very empty (*Voyager I*, our most distant spacecraft hurtling along at more than thirty-eight thousand miles per hour, will not reach the distance of even our sun's nearest neighbor, the Alpha Centauri system that it is not even headed toward, for more than seventy-five thousand years). Ergo, the probability of an ETI who is only slightly more advanced than we and also makes contact is virtually nil. If we ever do find ETI it will be as if a million-year-old *Homo erectus* were dropped into the middle of Manhattan, given a computer and cell phone, and instructed to communicate with us. ETI would be to us as we would be to this early hominid—godlike.

Science and technology have changed our world more in the past century than it changed in the previous hundred centuries. It took ten thousand years to get from the cart to the airplane, but only sixty-six years to get from powered flight to a lunar landing. Moore's Law of computer power doubling every eighteen months continues unabated and is now down to about a year. Ray Kurzweil, in *The Age of Spiritual Machines*, calculates that there have been thirty-two doublings since World War II, and that the Singularity point may be upon us as early as 2030. The Singularity (as in the center of a black hole, where matter is so dense that its gravity is infinite) is the point at which total computational power will rise to levels that are so far beyond anything that we can imagine that they will appear near infinite and thus, relatively speaking, be indistinguishable from omniscience.

When this happens the world will change more in a decade than it did in the previous thousand decades. Extrapolate that out a hundred thousand years, or a million years (an eye blink on an evolutionary time scale and thus a realistic estimate of how far advanced ETI will be, unless we happen to be the first spacefaring species, which is unlikely), and we get a gut-wrenching, mind-warping feel for just how godlike these creatures would seem.

In Clarke's 1953 novel *Childhood's End*, humanity reaches something like a Singularity (with help from ETIs) and must make the transition to a higher state of consciousness to grow out of childhood. One character early in the novel opines that "science can destroy religion by ignoring it as well as by disproving its tenets. No one ever demonstrated, so far as I am aware, the nonexistence of Zeus or Thor, but they have few followers now."

Although science has not even remotely destroyed religion, Shermer's Last Law predicts that the relationship between the two will be profoundly affected by contact with ETI. To find out how, we must follow Clarke's Second Law, venturing courageously past the limits of the possible and into the unknown.

36

Why ET Has Not Phoned In

The lifetime of civilizations in the Drake equation for estimating extraterrestrial intelligences is greatly exaggerated

In science there is arguably no more suppositional formula than that proposed in 1961 by radio astronomer Frank Drake for estimating the number of technological civilizations that reside in our galaxy: $N = Rf_p n_e f_l f_i f_c L$.

In this equation, N is the number of communicative civilizations, R is the rate of formation of suitable stars, f_p is the fraction of those stars with planets, n_e is the number of Earth-like planets per solar system, f_l is the fraction of planets with life, f_i is the fraction of planets with intelligent life, f_c is the fraction of planets with communicating technology, and L is the lifetime of communicating civilizations.

Although we have a fairly good idea of the rate of stellar formation ($R = 10$ sun-like stars per year is commonly accepted among astronomers), and we are confident that a significant number of these stars have planets, it is too soon to tell if any of them have Earth-like planets because the technology is not yet available to detect planets smaller than Jupiter-size behemoths. As for the rest of the equation's components, a dearth of data means that most SETI calculations are reduced to the creative speculations of quixotic astronomers.

Although most SETI astronomers are realistic about the limitations of their science, I was puzzled to encounter numerous caveats about *L*, the "lifetime" of technological civilizations, such as this one from SETI Institute astronomer Seth Shostak: "the lack of precision in determining these parameters pales in comparison to our ignorance of *L*." Similarly, the Mars Society president and space exploration visionary Robert Zubrin says that "the biggest uncertainty revolves around the value of *L*; we have very little data to estimate this number and the value we pick for it strongly influences the results of the calculation." Estimates of *L* reflect this uncertainty, ranging from 10 years to 10 million years, with a mean of about 50,000 years.

Using a conservative Drake equation calculation where $L=50{,}000$ years (and $R=10, f_p=0.5, n_e=0.2, f_l=0.2, f_i=0.2, f_c=0.2$), $N=400$ civilizations, or one per 4,300 light-years. Using Robert Zubrin's optimistic (and modified) Drake equation where $L=50{,}000$ years, $N=5{,}000{,}000$ galactic civilizations, or one per 185 light-years. (Zubrin's calculation assumes 10 percent of all 400 billion stars are suitable G- and K-type stars not part of multiples, with almost all having planets, and 10 percent of these containing an active biosphere, and 50 percent of those as old as Earth.) Estimates of *N* range wildly between these figures, from Planetary Society SETI scientist Thomas R. McDonough's 4,000 galactic civilizations to Carl Sagan's 1 million ETIs.

I find this inconsistency in the estimation of *L* perplexing because it is the one component in the Drake equation for which we have copious empirical data from the history of civilization on Earth. To compute my own value of *L* I compiled the lengths of 60 civilizations (the number of years from inception to demise), including Sumeria, Mesopotamia, Babylonia, the eight dynasties of Egypt, the six civilizations of Greece, the Roman Republic and Empire, and others in the ancient world, plus various civilizations since the fall of Rome, including the nine dynasties (and two republics) of China, four in Africa, three in India, two in Japan, six in Central and South America, and six modern states of Europe and America.

For all 60 civilizations in my database there was a total of 25,234 years, or $L=420.5$ years. For more modern and technological societies *L* became shorter, with the 28 civilizations since the fall of Rome averaging only 304.5 years. Plugging these figures into the Drake equation goes a long way toward explaining why ET has yet to drop by or phone in. Where $L = 420.56$ years, $N = 3.35$ civilizations in our galaxy; where $L = 304.53$ years, $N = 2.44$ civilizations in our galaxy. No wonder the galactic airways have been so quiet!

Although I am an unalloyed enthusiast for the SETI program, history tells us that civilizations may rise and fall in cycles too brief to allow enough to flourish at any one time to traverse (or communicate across) the vast and empty expanses between the stars. We evolved in small hunter-gatherer communities of 100 to 200 individuals; it may be that our species, and perhaps extraterrestrial species as well (assuming evolution operates in like manner elsewhere), are simply not well equipped to survive long periods in large populations.

Whatever the quantity of L, and whether N is less than 10 or more than 10 million, since we do not know for certain if it is more than 1 we need to ensure that L does not fall to 0 on our planet, the only source of civilization we have ever known.

37

The Chronology Conjecture Projector

Time machines, extraterrestrials, and the paradoxes of causality

In the original *Star Trek* series, Dr. McCoy falls through a time portal in a city "on the edge of forever" and changes the past in a way that erases the *Enterprise* and its crew, with the exception of Captain Kirk and Mr. Spock, who must return to the past to fix what McCoy has undone. Time travel is a well-worn staple of science fiction writers, but not only does it violate numerous physical laws, there are also fundamental problems of consistency and causality. The most prominent is the "matricide paradox" in which you travel back in time and kill your mother before she had you, which means you could not have been born to then travel back in time to kill your mother. In *Back to the Future*, Marty McFly faces a related but opposite dilemma in which he must arrange for his mother to date his father in order to ensure his conception.

One way around such paradoxes can be found in extremely sophisticated virtual reality machines (think of a holodeck), programmed to replicate a past time and place in such detail that it is indistinguishable from a real past (which one can never know in full in any case). Another option involves a multiple universes model of cosmology in which you travel back in time to a different but closely parallel universe to our own, as portrayed in Michael Crichton's novel *Timeline*, where the characters journey to another universe's medieval Europe without worry of mucking up our own chronology.

The fundamental shortcoming for both of these time travel scenarios is that it isn't *really* your past. A virtual reality time machine is simply a museum writ large, and transporting to some other universe's past would be like going back and meeting someone like your mother, who marries someone like your father, producing someone like, but not, you—surely a less appealing trip than one in your own timeline. To make that trip you need the time machine of Caltech's Kip Thorne, who had his interest piqued in time travel when he received a phone call one day from Carl Sagan, who was looking for a way to get the heroine of his novel *Contact*—Eleanor Arroway (played by Jodie Foster in the film version)—to the star Vega, 26 light-years away.

The problem Sagan faced, as all science fiction writers do in such situations, is that at the speed of, say, the *Voyager* spacecraft, it would take about 490,000 years to get to Vega. That's a long time to sit, even if you are in first class with your seat back and tray table down. Thorne's solution, adopted by Sagan, was to send Ellie through a wormhole—a hypothetical space warp similar to a black hole in which you enter the mouth, fall through a short tube in hyperspace that leads to an exit hole somewhere else in the universe (think of a tube running through the middle of a basketball—instead of going all the way around the surface of the ball to get to the other side, you tunnel through the middle). Since, as Einstein showed, space and time are intimately entangled, Thorne theorized that by warping space one might also be warping time, and that by falling through a wormhole in one direction it might be possible to travel backward in time.

Thorne's initial calculations showed that it was theoretically possible for Ellie to travel just one kilometer down the wormhole tunnel and emerge near Vega moments later—not even time for a bag of peanuts. After publishing his theory in a technical physics journal in 1988, the media got ahold of the story and branded Kip as "The Man Who Invented Time Travel." Not one to encourage such sensationalism, Thorne continued his research and by the early 1990s began growing skeptical of his own thesis.

Whether it is possible to actually travel through a wormhole without being crushed out of existence, Thorne reasoned, depends on the laws of quantum gravity, which are not fully understood at this point. What he and his colleagues ultimately discovered is that, as Kip told me, "all time machines are likely to self-destruct the moment they are activated." Thorne's colleague Stephen Hawking agreed, only half sardonically calling this conclusion the "chronology protection conjecture," in which "the laws of physics do not allow time machines," thus keeping "the world safe for historians." Besides, Hawking wondered, if time travel were possible, where are all the time tourists from the future?

It's a good question and, in conjunction with the paradoxes and physical law constraints, makes me skeptical as well. Until much more is known about quantum gravity and wormholes, virtual reality machines and multiple universes, I'll do my time traveling through the chronology projector of the mind.

38

Abducted!

Imaginary traumas are as terrifying as the real thing

In the wee hours of the morning of August 8, 1983, while I was traveling along a lonely rural highway approaching Haigler, Nebraska, a large craft with bright lights overtook me and forced me to the side of the road. Alien beings exited the craft and abducted me for ninety minutes, after which I found myself back on the road with no memory of what transpired inside the ship. The experience was real and I can prove that it happened because I recounted it to a film crew shortly after, and I am still in contact with some of the aliens.

The reality of my experience, however, is a separate question from what that experience represents. When alien abductees recount to me their stories, I do not deny that they had a real experience. And I believe that *they believe* their experience involved real aliens. But thanks to recent research by Harvard University psychologists Richard J. McNally and Susan A. Clancy, we now know that some fantasies are indistinguishable from reality, and they can be just as traumatic. In a 2004 paper in *Psychological Science*, "Psychophysiological Responding During Script-Driven Imagery in People Reporting Abduction by Space Aliens," McNally, Clancy, and their colleagues report the results of a study in which they measured heart rate, skin conductance, and left lateral frontalis electromyographic responses of claimed abductees as they relived their experiences through script-driven imagery. "Relative to control participants," the authors concluded, "abductees exhibited greater psychophysiological reactivity to abduction and stressful scripts than to positive and neutral scripts." In fact, the abductees' responses were comparable to those of post-traumatic stress disorder (PTSD) patients to scripts of their actual traumatic experiences.

The abduction study was initiated as a control in a larger investigation of memories of sexual abuse. In his 2003 book *Remembering Trauma* (Harvard University Press), McNally tracks the history of the recovered memory movement of the 1990s, in which some people, in an attempt (usually through hypnosis and guided imagery) to recover lost memories of childhood sexual molestation, in fact created false memories of abuse that never happened. "The fact that people who believe they have been abducted by space aliens respond like PTSD patients to audiotaped scripts describing their alleged abductions," McNally explains, "underscores the power of belief to drive a physiology consistent with actual traumatic experience." The vividness of a traumatic memory cannot be taken as evidence of its authenticity.

In the case of abductees, the most likely explanation is sleep paralysis and hypnopompic (upon awakening) hallucinations, in which flashing lights, buzzing sounds, and tingling sensations are accompanied by temporary paralysis and sexual fantasies, all of which are interpreted within a cultural context of UFOs and aliens so prevalent in today's pop culture. In addition, McNally found that abductees "were much more prone to exhibit false recall and false recognition in the laboratory than were control subjects" and they scored significantly higher than normal on a questionnaire measuring "absorption," a trait related to fantasy proneness that also predicts false recall.

My abduction experience was triggered by extreme sleep deprivation and physical exhaustion. I had just ridden a bicycle 83 straight hours and 1,259 miles in the opening days of the 3,100-mile nonstop transcontinental Race Across America. I was sleepily weaving down the road when my support motorhome flashed its brights and pulled alongside while my crew entreated me to take a sleep break. At that moment a distant memory of the 1960s television series *The Invaders* was inculcated into my waking dream. In the series, alien beings were taking over the Earth by replicating actual people but, inexplicably, retained a stiff little finger. Suddenly my support team was transmogrified into aliens. I stared intensely at their fingers, grilled my mechanic on bike technology, and interrogated my girlfriend on intimacies no alien would know.

After my 90-minute sleep break the experience represented nothing more than a bizarre hallucination, which I recounted to *ABC's Wide World of Sports* television crew filming the race. But at the time the experience was real, and that's the point. The human capacity for self-delusion is boundless, and the effects of belief are overpowering. Thanks to science we have learned to tell the difference between fantasy and reality.

VI

BORDERLANDS SCIENCE AND ALTERNATIVE MEDICINE

39

Nano Nonsense and Cryonics

True believers seek redemption from the sin of death

Timothy Leary's dead.

"No . . . he's outside looking in," sang the Moody Blues in their haunting 1960s ballad in words that would prove prophetic—after his 1996 death, seven grams of Leary's ashes were launched into orbit in a nine-by-twelve-inch canister, where they circled the Earth before burning up in a fiery finale befitting the man who had spent most of his life tripping out of this world by looking inward.

But according to documentary producer Paul Davids, whose graphic film *Timothy Leary's Dead* ends with a gruesome scene of Leary's head being hacked off, he isn't dead at all. He is cryonically frozen, awaiting reanimation. The Leary family emphatically denies it and Davids isn't talking. Both cryonics companies that Leary contacted in his final years—Alcor and Cryocare—assure me that Leary has gone the way of all flesh. No matter, since even cryonics proponents admit that anyone frozen to date will never be reanimated, unless . . .

The problem is obvious to anyone who has thawed a can of frozen strawberries. When they are frozen, the water within each cell expands, crystallizes, and shatters the cell wall. The overall structure remains intact while frozen, but when defrosted all the intracellular goo oozes out, turning your strawberries into runny mush. This is your brain on cryonics.

If even cryonicists recognize this detriment to the "suspension" of their "patients" (as they say in their optimistically worded lexicon), why would anyone bother spending $120,000 for a full-body freeze or $50,000 for just the "neural unit" (the head, to be reattached later to a cloned body), even if payment can be arranged through an insurance policy with the cryonics company as the beneficiary? The answer is nanotechnology. Microscopic machines with on-board computers will be injected into the defrosting corpse—er, I mean patient—programmed to repair the body molecule by molecule, cell by cell, until the trillions of cells are restored and the patient can be resuscitated. "Freeze—Wait—Reanimate" is the catchslogan of this scientistic religion in which nanotechnology will wash away the sin of death. The Resurrection is real . . . for all of us.

Every religion needs its gods, and cryonics has a trinity in Robert Ettinger (*The Prospect of Immortality*), Eric Drexler (*Engines of Creation*), and Ralph Merkle, whose magnum opus "The Molecular Repair of the Brain" can be downloaded at www.merkle.com. These works include just enough empirical data and logical reasoning to give one pause. This is a feasibility study premised on the fact that if you are cremated or buried there is zero probability of being resurrected. It is a secular version of Pascal's wager on God. Since the alternative is everlasting nothingness, the nano-cryonics scenario is worth the gamble.

Is it? That depends on how much time, effort, and money you are willing to invest in a program that has only a slightly higher probability than zero of succeeding. To believe in it takes a certain amount of faith in the secular religion of scientism—a blindly optimistic belief in the illimitable power of science to solve any and all problems, including death. Look how far we've come in just a century, believers argue, from the Wright brothers to Neil Armstrong in only sixty-six years. Average life expectancy has doubled, devastating diseases have been eradicated, and Moore's Law—the doubling of computer power every eighteen months (it's now down to about twelve)—continues unabated. Extrapolate these trend lines out a thousand years, or ten thousand, and immortality is virtually certain.

I want to believe the nano-cryonicists. Really I do. I gave up on religion in college but I often catch myself slipping back into my former evangelical fervor, now directed toward the wonders of science and nature. But this is precisely why I'm skeptical. It is too much like religion: it promises everything, delivers nothing (but hope), and is almost entirely based on faith. And if Ettinger, Drexler, and Merkle are the trinity of this scientistic sect, F. M. Esfandiary is its Saul who, on the road to his personal Damascus, became Paul when he changed his name to FM-2030 (his hundredth birthday and the year nano-cryonics is predicted to succeed) and declared, "I have no age. Am born and reborn every day. I intend to live forever. Barring an accident I probably will." He forgot about cancer, a pancreatic form of which killed him on July 10, 2000, three decades shy of immortality.

FM-2030—or more precisely his head—now resides in a vat of liquid nitrogen at the Alcor Life Extension Foundation in Scottsdale, Arizona, but his legacy lives on among his fellow "transhumanists" (they have moved beyond human) and "extropians" (they are against entropy), as his apostles Max More, Tom Morrow, and others who have reinvented themselves eagerly spread the meme of this branch of the church of scientism.

Is this science? No. Is it pseudoscience? No. It is what I call borderlands science—that fuzzy area between where scientistic-based claims reside that have yet to pass any tests but have some basis, however remote, in reality. It is not impossible for cryonics to succeed; it is just exceptionally unlikely (and new techniques are routinely developed, the latest being "vitrification," where the brain is hardened into a glass-like substance that avoids freezing damage).

Here we are faced with finding that exquisite balance between being credulous enough to accept a radical new idea that may turn out to be right, and skeptical enough not to be hoodwinked into believing bunkum. My credulity module is glad that at least a few scientists are devoting their careers to solving the problem of mortality; my skeptical module, however, indicates that transhumanistic-extropian nano-cryonics borders uncomfortably close to religion, and as such I worry, as Matthew Arnold did in his 1852 poem "Empedocles on Etna," that we will "feign a bliss of doubtful future date, And while we dream on this, Lose all our present state, And relegate to worlds . . . yet distant our repose."

Update: Cryonicist Ralph Merkle and I have maintained a correspondence since this essay was first published as he has tried to convince me to be more open-minded on the possibilities of cryopreservation. He points out, for example: "Your brain is an ordinary physical object that follows ordinary physical laws. There's no particular reason to believe that either making or repairing your brain violates any laws of physics, or is even all that fundamentally difficult. Your brain is just atoms, arranged in a particular way. Once we learn how to arrange atoms, we'll be able to repair cryopreserved brains." Maybe, but maybe not. Merkle's point is that one's "self" is stored in memory, and he points out: "There is no reason to believe that human long term memory is obliterated beyond recovery by modern cryopreservation methods. We have evidence that modern cryopreservation methods do, in fact, provide a quality of cryopreservation that is more than enough to preserve human long term memory in the information theoretic sense. The presence or absence of a synapse, as well as the proteins associated with the pre- and post-synaptic structures, and the proteins present in the synaptic cleft, should all be inferable following cryopreservation by today's methods." I remain skeptical that recovery of memory after cryopreservation will happen, but Merkle's point that, in principle, it is not impossible, is a reasonable one. As he notes: "The basic facts are simple. The human brain is physical, and human long term memory is associated with physical changes whose presence would still be identifiable following cryopreservation. Tissue stored at the temperature of liquid nitrogen remains essentially unchanged for at least many centuries. Computational power will increase enormously in the future, as will our ability to image and analyze the changes that have occurred in the cryopreserved human brain. Given your cryopreserved brain, and sufficient computational power, and a sufficient imaging technology, we will be able to recover the information that defines who you are. We will also be able to restore the cryopreserved human brain to a fully functional state, but strictly speaking this is not required for cryonics to work." If you want to read more go to www.merkle.com

40

I, Clone

The Three Laws of Cloning will protect clones and advance science

In his 1950 science fiction novel *I, Robot*, Isaac Asimov presented the Three Laws of Robotics: "1. A robot may not injure a human being, or, through inaction, allow a human being to come to harm. 2. A robot must obey the orders given it by human beings except where such orders would conflict with the First Law. 3. A robot must protect its own existence as long as such protection does not conflict with the First or Second Law."

Given the irrational fears people express today about cloning that parallel those surrounding robotics half a century ago, I would like to propose the "Three Laws of Cloning" that also clarify three misunderstandings about cloning: 1. A human clone is a human being no less unique in its personhood than an identical twin. 2. A human clone is a human being with all the rights and privileges that accompany this legal and moral status. 3. A human clone is a human being to be accorded the dignity and respect due any member of our species.

While such simplifications risk erasing the rich nuances found in ethical debates over pioneering science and technology, they do aid in attenuating risible fears often associated with such advances. Although it appears that the UFO Raelians have not succeeded in xeroxing themselves, it seems clear to most experts that someone, somewhere, some time soon is going to generate a human clone. And once one team has succeeded, it will be Katy-bar-the-door for others to bring on the clones.

If cloning produces genetic monstrosities that render it impractical as another form of fertility enhancement, then it will not be necessary to ban it because no one will use it. If cloning does work there is no reason to ban it because the three common reasons given for implementing restrictions are myths: the *Identical Personhood Myth*, the *Playing God Myth*, and the *Human Rights and Dignity Myth*.

The *Identical Personhood Myth* is well represented by Jeremy Rifkin: "It's a horrendous crime to make a Xerox of someone. You're putting a human into a genetic straitjacket." *Baloney*. These cloning critics have the argument bass ackward. Because they tend to be environmental determinists they should be arguing: "clone all you like, you'll never produce another you because environment matters as much as heredity." The best scientific evidence to date indicates that roughly half the variance between us is accounted for by genetics, the rest by environment. Because it is impossible to duplicate the near-infinite number of environmental permutations that go into producing an individual human being, cloning is no threat to unique personhood.

The *Playing God Myth* has numerous promoters, the latest being Stanley M. Hauerwas, a professor of theological ethics at Duke University, who responded to the Raelian cloning endeavor with this unequivocal denouncement: "The very attempt to clone a human being is evil. The assumption that we must do what we can do is fueled by the Promethean desire to be our own creators." In support of this myth he is not alone. A 1997 *Time*/CNN poll, conducted on the heels of the cloned sheep Dolly, revealed that 74 percent of Americans answered "yes" to the question "Is it against God's will to clone human beings?" *Balderdash*. Cloning may seem to be "playing God" only because it is unfamiliar. Consider earlier examples of once "godlike" fertility technologies that are now cheerfully embraced because we have become accustomed to them, such as in vitro fertilization, embryo transfer, and other fully sanctioned birth enhancement technologies.

The *Human Rights and Dignity Myth* is embodied in the Roman Catholic Church's official statement against cloning, based on the belief that it denies "the dignity of human procreation and of the conjugal union," as well as in a Sunni Muslim cleric's demand that "science must be regulated by firm laws to preserve humanity and its dignity." Members of Congress, assigned to deal more with legalities than moralities, have decreed that cloning violates the rights of the unborn. *Bunkum*. Clones will be no more alike than twins raised in separate environments, and no one is suggesting that twins do not have rights or dignity, or that twinning should be banned.

Instead of restricting or banning cloning, I propose that we adopt the Three Laws of Cloning, the principles of which are already incorporated in the laws and language of the U.S. Constitution, and allow science and medical technology to run their course. The soul of science is found in courageous thought and creative experiment, not in restrictive fear and prohibitions. For science to progress it must be given the opportunity to succeed or fail. Let's run the cloning experiment and see what happens.

41

Bottled Twaddle

Is bottled water tapped out?

In 1979 I started drinking bottled water. My bottles, however, contained tap water and were nestled in small cages on the frame of my racing bicycle.

Tap water was good enough then because we did not know how much healthier and tastier bottled water is. It must be, because Americans today spend more than $4 billion a year on it, paying 240 to 10,000 times more per gallon for bottled water than for tap. Bottled prices range from 70 cents to $5 per gallon, versus tap prices that vary from 45 cents to $2.85 per *thousand* gallons. We would not invest that amount of hard-earned sweat for nothing, would we? Apparently we would.

In March 1999, the Natural Resources Defense Council (NRDC) published the results of an extensive four-year study in which they tested more than 1,000 samples of 103 brands of bottled water, finding that "25 percent or more of bottled water is really just tap water in a bottle—sometimes treated, sometimes not." If the label says "from a municipal source" or "from a community water system," it's tap water.

Even more disturbing, the NRDC found that 18 of the 103 brands tested had, in at least one sample, "more bacteria than allowed under microbiological-purity guidelines." About one fifth of the waters "contained synthetic organic chemicals—such as industrial chemicals (e.g., toluene or xylene) or chemicals used in manufacturing plastic (e.g., phthalate, adipate, or styrene)," but these were "generally at levels below state and federal standards." The International Bottled Water Association issued a response to the NRDC study in which they state, "Close scrutiny of the water quality standards for chemical contaminants reveals that FDA bottled water quality standards are the same as EPA's tap water standards." Well, that's a relief, but in paying exceptional prices one might hope for exceptional quality.

One problem is that the FDA assigns less than one full-time employee to enforce compliance with bottled water regulations, which do not even apply to waters packaged and sold within the same state (about 65 percent of all sales), and are subject to less rigorous and frequent testing and purity standards for bacteria and chemical contaminants than those required of tap water. For example, bottled water plants must test for coliform bacteria once a week; city tap water must be tested a hundred or more times a month.

Spring water may call forth images of clean glacial runoff, but for some bottlers it is simply another season when their water is pumped out of wells and filtered, which FDA rules allow. Alaskan Falls water, for example, is bottled in Worthington, Ohio; Everest water springs from the municipal water supply of Corpus Christi, Texas.

If bottled water is not safer (a 2001 World Wildlife Fund study corroborated the general findings of the NRDC), then surely it tastes better. It does . . . as long as you believe in your brand. Enter the water wars hype. Pepsi preempted Coca-Cola with its blue-labeled Aquafina, so Coke countered with Dasani, a brand that includes a "Wellness Team" (meet Susie, Jonny, and Ellie, the "stress relief facilitator," "fitness trainer," and "lifestyle counselor," respectively, gleefully imbibing on the Dasani web page). Both charge more for their water than for their sugar water.

If the test is blind, however, the hype falls on deaf taste buds. In May 2001 ABC's *Good Morning America* found viewers' preferences to be Evian (12 percent), O-2 (19 percent), Poland Spring (24 percent), and good old New York City tap (45 percent). In July 2001 the *Cincinnati Enquirer* discovered that on a 1-to-10 scale the city's tap was rated at 8.2, compared with Dannon Spring Water's 8.3 and Evian Spring Water's 7.2. In 2001 the

Yorkshire water industry found that 60 percent of the 2,800 people tested could not tell the difference between its own tap and the United Kingdom's most popular bottled waters.

The most telling taste test was conducted by Showtime's television series *Penn and Teller's Bullshit*. They began with a blind comparison in which 75 percent of New Yorkers preferred city tap to the most expensive bottled waters. They then went to the left Coast and set up a hidden camera at a trendy Southern California restaurant that featured a water steward who dispensed elegant water menus to the patrons. All bottles were filled out of the same hose in the back; nevertheless, Angelenos were willing to plunk down $7 a bottle for L'eau du Robinet (French for "Faucet Water"), Agua de Culo (Spanish for "Ass Water"), Mt. Fuji (with "natural diuretics and antitoxins"), and Amazone ("filtered through the rainforest's natural filtration system"), declaring them all to be far superior in taste to tap water. There's no accounting for taste!

Bottled water does have one advantage over tap—you can take it with you wherever you go. So why not buy one bottle of each desirable size and refill it with your city's finest unnaturally filtered but salubriously delicious tap water?

42

Quantum Quackery

A surprise-hit film has renewed interest in applying quantum mechanics to consciousness, spirituality, and human potential

In spring 2004, I appeared on KATU-TV's *AM Northwest* in Portland, Oregon, with the producers of an improbably named film, *What the #@*! Do We Know?!* Artfully edited and featuring actress Marlee Matlin as a dreamy-eyed photographer trying to make sense of an apparently senseless universe, the film's central tenet is that we create our own reality through consciousness and quantum mechanics. I never imagined that such a film would succeed, but it has grossed millions and created a cult following.

The film's avatars are New Age scientists whose jargon-laden sound bites amount to little more than what Caltech physicist and Nobel laureate Murray Gell-Mann once described as "quantum flapdoodle." University of Oregon quantum physicist Amit Goswami, for example, says, "The material world around us is nothing but possible movements of consciousness. I am choosing moment by moment my experience. Heisenberg said atoms are not things, only tendencies." Okay, Amit, I challenge you to leap out of a twenty-story building and consciously choose the experience of passing safely through the ground's tendencies.

The work of Japanese researcher Masura Emoto, author of *The Message of Water*, is featured to show how thoughts change the structure of ice crystals—beautiful crystals form in a glass of water with the word "love" taped to it, whereas playing Elvis's "Heartbreak Hotel" causes a crystal to split into two. Would his "Burnin' Love" boil water?

The film's nadir is an interview with "Ramtha," a thirty-five-thousand-year-old spirit channeled by a fifty-eight-year-old woman named J. Z. Knight. I wondered where humans spoke English with an Indian accent thirty-five thousand years ago. Many of the films' producers, writers, and actors are members of Ramtha's "School of Enlightenment," where New Age pabulum is dispensed in costly weekend retreats.

The attempt to link the weirdness of the quantum world (such as Heisenberg's uncertainty principle, which states that the more precisely you know a particle's position, the less precisely you know its speed, and vice versa) to mysteries of the macro world (such as consciousness) is not new. The best candidate to link the two comes from physicist Roger Penrose and physician Stuart Hameroff, whose theory of quantum consciousness has generated much heat but little light in scientific circles.

Inside our neurons are tiny hollow microtubules that act like structural scaffolding. The conjecture (and that's all it is) is that something inside the microtubules may initiate a wave function collapse that leads to the quantum coherence of atoms, causing neurotransmitters to be released into the synapses between neurons and thus triggering them to fire in a uniform pattern, thereby creating thought and consciousness. Since a wave function collapse can only come about when an atom is "observed" (i.e., affected in any way by something else), neuroscientist Sir John Eccles, another proponent of the idea, even suggests that "mind" may be the observer in a recursive loop from atoms to molecules to neurons to thought to consciousness to mind to atoms. . . .

In reality, the gap between subatomic quantum effects and large-scale macro systems is too large to bridge. In his book *The Unconscious Quantum* (Prometheus Books) the University of Colorado physicist Victor Stenger demonstrates that for a system to be described quantum mechanically the system's typical mass m, speed v, and distance d must be on the order of Planck's constant h. "If mvd is much greater than h, then the system probably can be treated classically." Stenger computes that the mass of neural transmitter molecules, and their speed across the distance of the synapse, are about three orders of magnitude too large for quantum effects to be influential. There is no micro-macro connection. Then what the #$*! is going on here?

Physics envy. The history of science is littered with the failed pipe dreams of alluring reductionist schemes to explain the inner workings of the mind—schemes increasingly set forth in the ambitious wake of Descartes' own famous attempt, some four hundred years ago, to reduce all mental functioning to the actions of swirling vortices of atoms, supposedly dancing their way to consciousness. These Cartesian dreams have typically swept over their proud sponsors with the sure sense of certainty that only physics provides, but such dreams have typically faded away just as quickly, based on the many complexities of biology that their authors have refused to face and address.

What such reductionist schemes generally tell us, in fact, is far more about the dreamers than the object of their dreams. We see, so clearly in such schemes, the hubris of certain undeniably smart thinkers who believe that they, and they alone, have finally solved the most complex of all evolved phenomena, namely, how the mind works. The best science of the day, however, has invariably been achieved at a much more modest level by those researchers who have wisely tempered their reductionist cravings with a strong dose of humility and a respect for the surprises that biology constantly introduces into all reductionist schemes. Biology envy.

Update: Stuart Hameroff has vociferously objected to my use of the word "quackery" in this column published in Scientific American *because of its association with his name. This was not my intent, as this column was primarily about the film in which he was featured and not his work per se (and, he added, he too was skeptical of most of what appeared in the film). I remain skeptical of his theory of consciousness as being grounded in the microtubules of neurons, but Hameroff has sent me links to studies that show quantum effects can happen at molecular-level scales (and thus could, in principle, influence neural activity). He suggests this paper as the latest research in support of his theory: Hameroff, S., Penrose, R. 2014. "Consciousness in the Universe: A review of the 'Orch OR' Theory."* Physics of Life

Reviews *11: 39–78 10.1016/j.plrev.2013.08.002 Nevertheless, since we still do not understand how molecular activity in neurons translates into consciousness, making the jump from quantum effects at even the molecular scale into thought processes and mental experience seems to me to be unwarranted by the data.*

43

Hope Springs Eternal

Can nutritional supplements, biotechnology, and nanotechnology help us live forever?

As a skeptic I am often asked my position on one of the most extraordinary claims ever made: immortality. "I'm for it, of course," is my wiseacre reply.

Unfortunately, every one of the approximately hundred billion humans who lived before us has died, so the trend line does not bode well. Unless you follow the trend line generated by Ray Kurzweil and Terry Grossman in *Fantastic Voyage: Live Long Enough to Live Forever* (Rodale, 2004): "the rate of technical progress is doubling every decade, and the capability (price, performance, capacity, and speed) of specific information technologies is doubling every year. Because of this exponential growth, the 21st century will equal 20,000 years of progress at today's rate of progress." Within a quarter century, say the authors, "nonbiological intelligence will match the range and subtlety of human intelligence," then "soar past it because of the continuing acceleration of information-based technologies, as well as the ability of machines to instantly share their knowledge." Biotechnologies such as designer drugs and genetic engineering will halt the aging process; nanotechnologies such as nanorobots will repair and replace cells, tissues, and organs (including brains), reversing the aging process and allowing us to live forever.

In the meantime, to make sure you don't go the way of all flesh before this secular second coming (2030 by their calculation), you should employ "Ray and Terry's Longevity Program," which includes 250 supplements a day and weekly rounds of biochemistry reprogramming through intravenous "nutritionals" and acupuncture. To boost antioxidant levels, for example, Kurzweil suggests a concoction of "alpha lipoic acid, coenzyme Q_{10}, grapeseed extract, resveratrol, bilberry extract, lycopene, silymarin, conjugated linoleic acid, lecithin, evening primrose oil (omega-6 essential fatty acids), n-acetyl-cysteine, ginger, garlic, 1-carnitine, pyridoxal-5-phosphate, and Echinacea." *Bon appétit.*

Ray Kurzweil is a brilliant and creative mind—the inventor of the first optical character recognition program and CCD flatbed scanner, inventor of the first print-to-speech reading machine for the blind and text-to-speech synthesizer, recipient of the 1999 National Medal of Technology, and inductee into the National Inventor Hall of Fame. His books *The Age of Intelligent Machines* and *The Age of Spiritual Machines* significantly influenced the field of artificial intelligence. Thus when Ray Kurzweil speaks, people listen. But my baloney detection alarm went off in three areas of his work.

One, I'm skeptical of the effectiveness of nutritional supplements. When I was bike racing in the 1980s I went through a period of megadosing vitamins and minerals that produced brightly colored urine but little else. The testimonials behind such nutritional claims are powerful but the science is weak. The fact that the field is fraught with fads and ever-changing claims for *X* as the elixir of health and longevity does not bode well. Nutritional science says that we get virtually all of the vitamins and minerals we need through a balanced diet, and that more is not better (see www.nutriwatch.org). While these diets are helping more of us live longer and healthier lives, they are not helping anyone live longer than the maximum human life span of about 120 years. The 56-year-old Kurzweil says that his program has reduced his biological age to about 40. I'm no aging expert or carny barker, but if I had to guess his age from his author photo I'd say, uh, 56.

Two, I'm skeptical of extrapolating trend lines very far into the future. Human history is highly nonlinear and unpredictable. Plus, in my opinion, the problems of aging and artificial intelligence are orders of magnitude harder than anyone anticipated. Machine intelligence of a human nature is probably a century away, and immortality is at least a millennium away, if not unattainable altogether.

Three, I'm skeptical whenever people argue that the Big Thing is going to happen in *their* lifetime. Evangelicals never claim that the Second Coming is going to happen in the *next* generation (or that they will be "left behind" while others are saved). Likewise, secular doomsayers typically predict the demise of civilization within their allotted time (but that they will be part of the small surviving enclave). Prognosticators of both religious and secular utopias always include themselves as members of the chosen few, and paradise is always within reach.

Hope springs eternal.

Update: I highly recommend the documentary film Transcendent Man, *about Ray Kurzweil and his efforts to achieve immortality: transcendent man.com*

44

Full of Holes

The curious case of acupuncture

John Marino was the most driven man I ever met, a monomaniac on a mission to break the transcontinental cycling record, which he did in 1980, covering the three thousand miles in twelve days, three hours. I wanted to be like John, so that year I took up serious cycling, and in addition to pedaling hundreds of miles a week with him, I followed his training regimen of vegetarian meals, megavitamin dosing, fasting, colonics, mud baths, iridology (iris reading), negative ions, chiropractic, massage, and acupuncture.

Although most of the nostrums I tried were useless, I noted with interest (because he beat me) that the winner of the 1985 Race Across America (cofounded by Marino and myself), Jonathan Boyer, had a Chinese acupuncturist on his support crew. Given the success of Marino and Boyer, it seemed possible that there might be a biomedical connection with acupuncture, even if the theory behind it is baloney.

Traditional Chinese medicine holds that a life energy called qi ("chee") flows through meridians in the body; each of the 12 main meridians represents a major organ system. On these 12 meridians are 365 acupuncture points, one for each day of the year. When yin and yang are out of balance, qi can become blocked, leading to illness. Inserting needles at the blocked points—today believed to number about 1,000—supposedly stimulates healing and health.

This theory lacks any basis in biological reality because nothing like qi has ever been found by science. Nevertheless, a medicinal procedure may work for some other reason not related to the original erroneous theory. Under certain very limited conditions, some forms of acupuncture may work. According to Dr. George A. Ulett, a practicing physician and acupuncturist (with both an MD and a PhD) and the author of the 1992 *Beyond Yin and Yang: How Acupuncture Really Works* and the 2002 textbook *The Biology of Acupuncture* (both published by Warren H. Green in St. Louis), electroacupuncture—the electrical stimulation of tissues through acupuncture needles—increases the effectiveness of analgesic (pain-relieving) acupuncture by as much as 100 percent over traditional acupuncture. Ulett posits that electroacupuncture stimulates the release of such neurochemicals as beta-endorphin, enkephalin, and dynorphin, leading to pain relief. In fact, says Ulett, the needles are not even needed; electrically stimulating the skin (transcutaneous nerve stimulation) is sufficient. Using this technique, Ulett cites research in which the amount of gas anesthetic used in surgery was reduced by 50 percent.

These findings might help explain the results of a study published in the May 4, 2005, issue of the *Journal of the American Medical Association*, in which Klaus Linde and his colleagues at the Technische Universitaet in Munich, Germany, compared the experiences of 302 people suffering from migraines who received either acupuncture, sham acupuncture (needles inserted at nonacupuncture points), or no acupuncture. During the study, the patients kept headache diaries. Subjects were "blind" to which experimental group they were in; the evaluators also did not know whose diary they were reading. Professional acupuncturists administered the acupuncture and sham acupuncture treatments. The results were dramatic: "The proportion of responders (reduction in headache days by at least 50%) was 51% in the acupuncture group, 53% in the sham acupuncture group, and 15% in the waiting list group." The authors concluded that this effect "may be due to nonspecific physiological effects of needling, to a powerful placebo effect, or to a combination of both."

I have had acupuncture treatments a number of times and can attest to the fact that while "needling" (where the acupuncturist taps and twists the flesh-embedded needle) isn't painful, it is most definitely noticeable. If acupuncture works beyond placebo, it is through the physical stimulation and release of the body's natural painkillers. Finding that sham acupuncture is as effective as "real" acupuncture demonstrates that the qi theory is full of holes. The effects of being poked by needles, however, cannot be ignored. Understanding the psychology and neurophysiology of acupuncture and pain will lead to a better theory. And for all such alternative medicine claims, testimonials can only steer us in the direction of where to conduct research; science is the only tool that can tell us whether they really work.

45

Airborne Baloney

The latest fad in cold remedies is full of hot air

> The first principle is that you must not fool yourself—and you are the easiest person to fool.
>
> —Richard Feynman, Caltech physicist and Nobel laureate

I violated Feynman's first principle during a recent book tour that took me daily through congested airports, on crowded jets, and in crammed bookstores amid sneezing, coughing, germ-infested multitudes. One day, while squeezed into the sardine section of coach, with the guy behind me obeying the command of the germs in his lungs to go forth and multiply, I cursed myself for having forgotten my Airborne tablets, an orange-flavored effervescent concoction of herbs, antioxidants, electrolytes, and amino acids that fizzles into action in a glass of water. You down it "at the first sign of a cold symptom or before entering crowded environments," says the package, most notably in "airplanes, restaurants, offices, hospitals, schools, health clubs, carpools, theaters, sports arenas."

Airborne was not my first foray into alternative medicine. Ever since I took up bike racing in 1980 I have been imbibing Emer'gen-C, packets of 1,000 mg powdered vitamin C, glucosamine, chondroitin, potassium, sodium, and other ingredients supposedly vital for energy and health. "Vitamin C is essential for the formation and maintenance of connective tissue, is a powerful antioxidant, and it is involved in normal immune function," says the package.

In the logic-tight compartments of my brain, my magic module trumped my skeptic module and I hadn't given this product any thought until, much to my chagrin, the host for one of my book tour stops, a Menlo Park Internet venture capitalist and science blogger named David Cowan, mentioned that he had debunked Airborne in a recent blog (http://whohastimeforthis.blogspot.com/). A science-savvy investor, Cowan was quick to spot the clever marketing technique of suggesting that Airborne prevents or cures colds without actually saying so. "Take at the FIRST sign of a cold symptom or before entering crowded environments," the instructions say. Then, "Repeat every three hours as necessary." In the (really) fine print, however, it says, "These statements have not been evaluated by the Food and Drug Administration. This product is not intended to diagnose, treat, cure, or prevent any disease." It is, in fact, classified as a dietary supplement.

Even more perfidious is how the company turned a liability into an asset. Most drugs are developed by Big Pharma—Brobdingnagian corporations with vast teams of scientists who have, to date, failed to cure the common cold. Airborne was created by "Knight McDowell Labs"—Victoria Knight is a schoolteacher and Rider McDowell is a scriptwriter. Instead of hiding their lack of credentials, they boast about them at the top of their web page (www.airbornehealth.com): CREATED BY A SECOND GRADE SCHOOL TEACHER! "As any confidence artist knows," Cowan explains, "disclosing unflattering facts up front wins the target's trust." And $100 million in annual sales is all the data the lab needs. Subsequently they invented Airborne Gummi Lozenges, Airborne Junior 3-Pack (half the strength of the original), and the Airborne Super Duper Combo Frequent Flyer pack. Kevin Costner says, "I wouldn't go on a movie set without it; it's on my plane and in my house." His plane? I guess he doesn't fly coach. No wonder he thinks Airborne is so effective.

As for real scientific data on Airborne, there is a web page used to provide a link to "clinical results" (no longer there), but when Cowan wrote the company for the data, he received this reply: "The 2003 trial was a small study conducted for what was then a small company. While it yielded very strong results, we feel that the methodology (protocol) employed is not consistent with our current product usage recommendations. Therefore, we no longer make it available to the public." Why? The company

CEO, Elise Donahue, told ABC News: "We found that it confused consumers. Consumers are really not scientifically minded enough to be able to understand a clinical study."

ABC News looked into the clinical trial and discovered that it was conducted by GNG Pharmaceutical Services, "a two-man operation started up just to do the Airborne study. There was no clinic, no scientists and no doctors. The man who ran things said he had lots of clinical trial experience. He added that he had a degree from Indiana University, but the school says he never graduated."

In one final lunge at product verisimilitude (dang it, that zesty taste feels like it works), I consulted Dr. Harriet Hall, a retired air force flight surgeon and family physician who studies alternative medicine. Hall looked up Airborne's ingredients in the Natural Medicines Comprehensive Database, "a trustworthy nonpartisan compilation of all the published research on natural medicines," and found no evidence that any of the ingredients prevents colds. "A combination of antioxidants including A, E, and C has been recommended for a number of conditions," Hall explained, "but it seems every new study shows a poorer outcome with antioxidants than without. Effects that can be demonstrated in the lab often don't carry over into improved patient outcomes. Human bodies are much more complex than test tubes!" Worse, vitamin A is unsafe in doses over 10,000 units a day, and Airborne contains 5,000 units per pill and recommends five pills a day or more. The only positive finding was for vitamin C, for which there is some evidence that taking high doses may shorten the duration of cold symptoms by 1 to 1.5 days in some patients, but that the large amounts needed may cause side effects. "There's more evidence for chicken soup than for Airborne," Hall told me. "In the absence of any credible double-blind studies to support the claims for Airborne, I'll stick to hand washing."

Chicken soup for the traveler's soul.

Update: In 2008 the Federal Trade Commission accused the makers of Airborne of false advertising for suggesting their product could help ward off bacteria and germs associated with common colds and the flu. Subsequently, *a class action lawsuit was filed against the company Airborne Health, Inc., claiming that Airborne falsely advertised the therapeutic components of its product. On March 4, 2008, the former owners of Airborne Health, Inc., agreed to pay $23.3 million to settle the lawsuit: money.cnn.com/2008/03/04/news/companies/airborne_settlement*

46

Eat, Drink, and Be Merry

Or why we should learn to stop worrying and love food

Among athletes who obsess about their weight, we cyclists are second to none. Training rides are filled with conversations about weight lost or gained and the latest diet regimens and food fads. Resolutions are made and broken. Guilt increases with each pound, accompanied by quiet relief that cycling shorts are black and made of stretchy Lycra. We all know the formula: ten pounds of extra weight on a 5 percent grade slows your ascent by half a mile an hour. It has a ring of Newtonian finality to it. $F=MA$. The force needed to turn the pedals equals acceleration times that mass on the saddle.

But most of the guys I ride with are like me: in their forties and fifties with jobs and families and long past racing prime. We ride because it's fun, and it feels good to be fit. So why obsess over a few pounds? Because that is the cycling culture—emblematic of our society at large—that carries its own internal calculus: $G=FT$. Guilt is proportional to the frequency and tastiness of the food.

The problem is that evolution designed our bodies to crave copious amounts of rich and tasty foods because in the Paleolithic Period such foods were valuable and rare. How can we resist? We shouldn't, says Barry Glassner, a University of Southern California sociologist and author of the

2007 book *The Gospel of Food: Everything You Think You Know About Food Is Wrong* (Ecco). We have wrongly embraced what Glassner calls "the gospel of naught," the view that "the worth of a meal lies principally in what it lacks. The less sugar, salt, fat, calories, carbs, preservatives, additives, or other suspect stuff, the better the meal." The science behind this culinary religion, says Glassner, is close to naught. Religion is the right descriptor, as the problem began with American Puritanism, or the obsessive fear that somewhere out there people are enjoying themselves and should be stopped. Yet, says Glassner, "Studies of dieters find that those who regard pleasure as unimportant in their food choices enjoy their meals less and are more likely to be dissatisfied with their bodies and exhibit symptoms of eating disorders."

Taste matters. Glassner cites a study in which "Swedish and Thai women were fed a Thai dish that the Swedes found overly spicy. The Thai women, who liked the dish, absorbed more iron from the meal. When the researchers reversed the experiment and served hamburger, potatoes, and beans, the Swedes, who like this food, absorbed more iron. Most telling was a third variation of the experiment, in which both the Swedes and the Thais were given food that was high in nutrients but consisted of a sticky, savorless paste. In this case, neither group absorbed much iron."

Speaking of iron, Atkins is out and meat is bad, right? Wrong. Glassner notes a study showing that as meat consumption and blood cholesterol levels increased in groups of Greeks, Italians, and Japanese, their death rates from heart disease decreased. Of course, many other variables are involved in determining causal relationships between diet and health. Glassner cites a study showing a 28 percent decreased risk of heart attacks among nonsmokers who exercised thirty minutes a day, consumed fish, fiber, and folate, and avoided saturated and trans fats and glucose-spiking carbs. And according to Harvard epidemiologist Karin Michels, "It appears more important to increase the number of healthy foods regularly consumed than to reduce the number of less healthy foods regularly consumed."

It's more complicated still. Glassner reviews research showing that heart disease, cancer, and other illnesses are significantly increased by "viral and bacterial infections, job stress, living in distressed neighborhoods, early deficits such as malnutrition, low birth weight, lack of parental support, and chronic sleep loss during adolescence and adulthood." Another study found that such diseases "are higher in states where participation in civic life is low, racial prejudice is high, or a large gap exists between the incomes of the rich and poor and of women and men."

In trying to make sense of this cornucopia of data Glassner quotes a former editor of the *New England Journal of Medicine*, Marcia Angell: "Although we would all like to believe that changes in diet or lifestyle can greatly improve our health, the likelihood is that, with a few exceptions such as smoking cessation, many if not most such changes will produce only small effects. And the effects may not be consistent. A diet that is harmful to one person may be consumed with impunity by another."

Perhaps here the wisdom of the preacher in Ecclesiastes 8:15 should be heeded: "Then I commended mirth, because a man hath no better thing under the sun, than to eat, and to drink, and to be merry."

VII

PSYCHOLOGY AND THE BRAIN

47

The Captain Kirk Principle

Intuition is the key to knowing without knowing how you know

Stardate: 1672.1. Earthdate: October 6, 1966. *Star Trek* episode 5, "The Enemy Within." Captain James T. Kirk has just beamed up from planet Alpha 177, where magnetic anomalies have caused the transporter to malfunction, splitting Kirk into two beings. One is cool and rational. The other is impulsive and irrational. Rational Kirk must make a command decision to save the crew, but he is paralyzed with indecision, bemoaning to Dr. McCoy: "I can't survive without him [irrational Kirk]. I don't want to take him back. He's like an animal—a thoughtless, brutal animal. And yet it's me."

This psychological battle between intellect and intuition was played out in nearly every episode of *Star Trek* in the characters of the ultrarational Mr. Spock and the hyperemotional Dr. McCoy, with Captain Kirk as the near-perfect embodiment of both. Thus I call this balance the *Captain Kirk Principle: intellect is driven by intuition, intuition is directed by intellect.*

For most scientists intuition is the bête noire of a rational life, the enemy within to beam away faster than a phaser on overload. Yet the Captain Kirk Principle is now finding support from a rich new field of scientific inquiry brilliantly summarized by Hope College psychologist David G. Myers in his new book *Intuition: Its Powers and Perils* (Yale

University Press, 2002). I confess to being skeptical when I first picked up the book, but as Myers demonstrates through countless well-replicated experiments, intuition—"our capacity for direct knowledge, for immediate insight without observation or reason"—is as much a part of our thinking as analytic logic.

Physical intuition, of course, is well known and accepted as part of an athlete's repertoire of talents—Michael Jordan and Tiger Woods come to mind. But there are social and psychological intuitions as well, which operate at levels so fast and subtle that they cannot be considered a function of rational thought. Harvard's Nalini Ambady and Robert Rosenthal, for example, discovered that the evaluation of teachers by students who saw a mere thirty-second video of the teacher were remarkably similar to those of students who had taken the course. Even three two-second video clips of the teacher yielded a striking 0.72 correlation with the course student evaluations.

Research consistently shows how unattended stimuli can subtly affect us. At USC, Moshe Bar and Irving Biederman flashed emotionally positive scenes (kitten, romantic couple) or negative scenes (werewolf, dead body) for forty-seven milliseconds before subjects viewed slides of people. Although subjects reported seeing only a flash of light for the initial emotionally charged scenes, they gave more positive ratings to people whose photos had been associated with the positive scenes—i.e., something registered somewhere in the brain.

Intuition similarly plays a role in "knowing" other people. The best predictor of how well a psychotherapist will work out for you is your initial reaction in the first five minutes of the first session. People with dating experience know within minutes whether they will want to see a first date again. To the extent that lie detection through the observation of body language and facial expressions is accurate (overall not very), women are better at it than men because they are more intuitively sensitive to subtle cues. Women are also superior in discerning which of two people in a photo was the other's supervisor, whether a male-female couple is a genuine romantic relationship or a posed, phony one, and when shown a two-second silent video clip of an upset woman's face, women guess more accurately than men whether she is criticizing someone or discussing her divorce.

Intuition is not subliminal perception; it is subtle perception and learning—knowing without knowing that you know. Chess masters often "know" the right move to make even if they cannot articulate how they know it. People who are highly skilled in identifying "micromomentary" facial expressions are also more accurate in judging lying. (In testing college students, psychiatrists, polygraphists, court judges, police officers, and Secret Service agents on their ability to detect lies, only Secret Service agents, trained to look for subtle cues, scored above chance.)

Most of us are not good at lie detection because we rely too heavily on what people say rather than on what they do. Subjects with damage to the brain that renders them less attentive to speech are more accurate at detecting lies, such as aphasic stroke victims who were able to identify liars 73 percent of the time when focusing on facial expressions (nonaphasic subjects did no better than chance). We may even be hard-wired for intuitive thinking: a patient with damage to parts of his frontal lobe and amygdala (the fear center) is unable to understand social relations or detect cheating, particularly in social relations, even though cognitively he is otherwise normal.

Although in science we eschew intuition because of its many perils (also thoroughly documented by Myers), we would do well to remember the Captain Kirk Principle that intellect and intuition are complementary, not competitive. Without intellect our intuition may drive us unchecked into emotional chaos. Without intuition we risk failing to resolve complex social dynamics and moral dilemmas, as Dr. McCoy explained to the indecisive rational Kirk: "We all have our darker side—we need it! It's half of what we are. It's not really ugly—it's human. Without the negative side you couldn't be the captain, and you know it! Your strength of command lies mostly in him."

48

None So Blind

Perceptual-blindness experiments challenge the validity of eyewitness testimony and the metaphor of memory as a video recording

Picture yourself watching a one-minute video of two teams of three players each, one team donning white shirts and the other black shirts, as they move about each other in a small room tossing two basketballs. Your task is to count the number of passes made by the white team. Unexpectedly, after thirty-five seconds a gorilla enters the room, walks directly through the farrago of bodies, thumps his chest, and nine seconds later exits (see figure). Would you see the gorilla?

Most of us, in our perceptual vainglory, believe we would—how could anyone miss a guy in an ape suit? In fact, 50 percent of subjects in this remarkable experiment by Daniel J. Simons and Christopher F. Chabris do not see the gorilla, even when asked if they noticed anything unusual (see their paper "Gorillas in Our Midst" at http://bit.ly/Z3I3Es with links to ordering the DVD of this and other clips of related experiments). The effect is known as "inattentional blindness": when attending to one task—say, talking on a cell phone while driving—many of us become blind to dynamic events, such as a gorilla in the crosswalk.

I incorporated the gorilla DVD into my public lecture on science and skepticism given at universities around the country. I always ask for a show of hands of those who did not see the gorilla in the first viewing (I show

the clip a second time with no counting, and nearly everyone sees it). Even under such public social pressure, out of more than ten thousand students I encountered last year (2003), approximately half confessed their perceptual blindness. Many were shocked, accusing me of showing two *different* clips. Simons had the same experience: "we actually rewound the videotape to make sure subjects knew we were showing them the same clip."

These experiments reveal a hubris in our powers of perception, as well as a fundamental misunderstanding of how the brain works. We think of our eyes as video cameras, and our brains as blank tapes to be filled with percepts. Memory, in this model, is simply rewinding the tape and playing it back in the theater of the mind, in which some cortical commander watches the show and reports to a higher homunculus what it saw.

Fortunately for criminal defense attorneys, this is not the case. The perceptual system, and the brain that analyzes its data, are far more complex. As a consequence, much of what passes before our eyes may be invisible to a brain focused on something else. "The mistaken belief that important events will automatically draw attention is exactly why these findings are surprising; it is also what gives them some practical implications," Simons told me. "By taking for granted that unexpected events will be seen, people often are not as vigilant as they could be in actively anticipating such events."

Driving is an example. "Many accident reports include claims like 'I looked right there and never saw them,'" Simons notes. "Motorcyclists and bicyclists are often the victims in such cases. One explanation is that car drivers expect other cars but not bikes, so even if they look right at the bike, they sometimes might not see it." Simons recounts a study by Richard Haines of pilots who were attempting to land a plane in a simulator with the critical flight information superimposed on the windshield.

"Under these conditions, some pilots failed to notice that a plane on the ground was blocking their path."

Over the years in this column I have pounded paranormalists pretty hard, so they may rightly point to these studies and accuse me of inattentional blindness when it comes to ESP and other perceptual ephemera. Perhaps my attention to what is known in science blinds me to the unknown.

Maybe. But the power of science lies in open peer publication, which, with the emergence of the Internet, is no longer constrained by the affordances of paper. I may be perceptually blind, but not all scientists will be, and out of this fact arises the possibility of new percepts and paradigms. There may be none so blind as those who will not see, but in science there are always those whose vision is not so constrained. But first they must convince the skeptics, and we are trained to look for gorillas in our midst.

49

Common Sense

Surprising new research shows that crowds are often smarter than individuals

In 2002 I was asked by an acquaintance to serve as his "phone a friend" on the popular television series *Who Wants to Be a Millionaire?* When he was stumped by a question he elected to "poll the audience" instead, which was wise not only because I did not know the answer, but also because the data show that the audience is right 91 percent of the time, compared to only 65 percent for experts.

Although this difference may in part be explained by the fact that the audience is usually queried for easier questions, there is something deeper at work here. For solving a surprisingly large and varied number of problems, crowds are smarter than individuals. This is contrary to what the nineteenth-century Scottish journalist Charles Mackay concluded in his 1841 book *Extraordinary Popular Delusions and the Madness of Crowds*, a staple of skeptical literature: "Men, it has been well said, think in herds. It will be seen that they go mad in herds, while they only recover their senses slowly, and one by one." This has been the dogma ever since, supported by sociologists such as Gustave Le Bon, in his classic work *The Crowd: A Study of the Popular Mind*. "In crowds it is stupidity and not mother wit that is accumulated."

Au contraire, Monsieur Le Bon. There is now overwhelming evidence, artfully accumulated and articulated by the *New Yorker* columnist James Surowiecki in his enthralling 2004 book *The Wisdom of Crowds* (Doubleday), that "the many are smarter than the few." In one experiment subjects were asked to estimate the number of jelly beans in a jar. The group average was 871, only 2.5 percent off the actual figure of 850. Only one of the fifty-six subjects was closer. The reason is that in a group individual errors on either side of the true figure cancel each other out.

A similar result was discovered in an example so counterintuitive that it startles. When the US submarine *Scorpion* disappeared in May 1968, a naval officer named John Craven assembled a diverse group of submarine experts, mathematicians, and salvage divers. Instead of putting them in a room to consult one another, he had each of them give a best guesstimate, based on the sub's last known speed and position (and nothing else), of the cause of its demise, its rate and steepness of descent, and other variables. Craven then computed a group average employing Bayes's theorem, a statistical method where a probability is assigned to each component of a problem (see chapter 71 on a Bayesian computation of the probability of God's existence). The *Scorpion*'s location on the ocean floor was only 220 yards from the averaged prediction even though not one member of the group had selected that spot.

Stranger still was the stock market's reaction on January 28, 1986, the day of the space shuttle *Challenger* catastrophe. Of the four major shuttle contractors—Lockheed, Rockwell International, Martin Marietta, and Morton Thiokol—the latter (the builder of the solid rocket booster that exploded) was hit hardest, with a 12 percent loss, compared to only 3 percent for the others. A detailed study of the market (a sizable crowd indeed!) by economists Michael T. Maloney from Clemson University and J. Harold Mulherin from Claremont McKenna College could find no evidence of insider trading or media focus on the rocket booster or Morton Thiokol. Given four possibilities, the masses voted correctly.

Not all crowds are wise, of course—lynch mobs come to mind. And "herding" can be a problem when the members of a group think uniformly in the wrong direction. The stock market erred for the space shuttle *Columbia* disaster, for example, dumping Thiokol stock even though the boosters were not involved.

For a group to be smart it should be autonomous, decentralized, and cognitively diverse, which the committee who rejected the foam impact theory of the space shuttle *Columbia* while it was still in flight was not. Google is brilliant because it uses an algorithm that ranks web pages by the number of links to them, with those links themselves valued by the number of links to their page of origin. This works because the Internet is the largest autonomous, decentralized, and diverse crowd in history, IMHO.

50

Murdercide

Science unravels the myth of suicide bombers

> You should be very proud of me. It's an honor, and you will see the results, and everybody will be happy. I want you to remain very strong as I knew you, but whatever you do, head high, with a goal, never be without a goal, always have a goal in front of you and always think, "what for."
>
> —Final letter to his wife by Ziad Jarrah, terrorist who piloted Flight 93 into a Pennsylvania field on September 11, 2001

Semantic precision is one of the hallmarks of science, which is why I think we need a new term for suicide bombings. "Suicide" is "taking one's own life, self-murder." The point of blowing yourself up is not to end your own life, but to take the lives of others. This is why it was so puzzling when the host of the ABC television series *Politically Incorrect,* Bill Maher, was fired when he opined five days after 9/11: "Staying in the airplane when it hits the building—say what you want about it, it's not cowardly." Maher is right. Since the *Oxford English Dictionary* defines a coward as "one who displays ignoble fear or want of courage in the face of danger, pain, or difficulty," it seems unlikely that this is the type of person a terrorist organization would solicit for blowing up buildings and people. Even *suicide bomber* is not the right term.

(Who said it was cowardly? President George W. Bush. Of course, he was just following presidential convention from both parties: President Clinton said the terrorists who bombed the US embassies in Nairobi, Kenya, and Dar-es-Salaam in 1998 were cowards, while President Reagan called the 1983 terrorist bombing of the US embassy in Beirut cowardly.)

Genocide—"the deliberate and systematic extermination of an ethnic or national group"—is not quite right either, because the goal of suicide bombings is to strike terror into the ethnic or national majority by killing a minority of them.

Police have an expression for people who put themselves into circumstances that force officers to shoot them: "suicide by cop." Following this lingo, suicide bombers commit "suicide by murder," so we might call such acts "murdercide": *the killing of a human or humans with malice aforethought by means of self-murder.*

The reason we need semantic precision is that suicide has drawn the attention of scientists who understand it to be the product of two conditions quite unrelated to murdercide: ineffectiveness and disconnectedness. According to the Florida State University psychologist Thomas Joiner, in his remarkably revealing scientific treatise *Why People Die by Suicide* (Harvard University Press, 2006): "People desire death when two fundamental needs are frustrated to the point of extinction; namely, the need to belong with or connect to others, and the need to feel effective with or to influence others." In other words, people commit suicide when they feel personally ineffective and socially disconnected, and have acquired the capability for serious self-harm and have habituated to the fear about pain involved in the act itself. These are necessary but not sufficient conditions for suicide—not everyone who has them kills themselves, but those who kill themselves have them.

By this theory, the people who chose to jump from the World Trade Center rather than burn to death were not suicidal; neither were the passengers on Flight 93 who courageously fought the hijackers for control of the plane that ultimately crashed into a Pennsylvania field; and neither were the hijackers who flew the planes into the buildings.

The belief that suicide bombers are poor, uneducated, disaffected, or disturbed is contradicted by science. Marc Sageman, a forensic psychiatrist, former CIA case officer in Afghanistan, and now a senior fellow at the Foreign Policy Research Institute, found in a study of four hundred Al Qaeda members: "three quarters of my sample came from the upper or middle class. The vast majority—90 percent—came from caring, intact families. Sixty-three percent had gone to college, as compared with the 5–6 percent that's usual for the third world. These are the best and brightest of their societies in many ways." Nor were they sans employment and familial duties. "Far from having no family or job responsibilities,

73 percent were married and the vast majority had children. Three-quarters were professionals or semi-professionals. They are engineers, architects, and civil engineers, mostly scientists. Very few humanities are represented, and quite surprisingly very few had any background in religion."

According to Joiner, a necessary condition for suicide is habituation to the fear about the pain involved in the act. How do terrorist organizations infuse this condition in their recruits? One way is through psychological reinforcement. According to University of Haifa political scientist Ami Pedahzur in *Suicide Terrorism* (Polity Press, 2005), the celebration and commemoration of suicide bombings that began in the 1980s changed the culture into one that idolized martyrdom and its heroes. Today, murderciders are posterized like star athletes.

Another method of control is "group dynamics," says Sageman: "The prospective terrorists joined the jihad through pre-existing social bonds with people who were already terrorists or had decided to join as a group. In 65 percent of the cases, pre-existing friendship bonds played an important role in this process." Those personal connections help override the natural inclination to avoid self-immolation. "The suicide bombers in Spain are another perfect example. Seven terrorists sharing an apartment and one saying 'Tonight we're all going to go, guys.' You can't betray your friends, and so you go along. Individually, they probably would not have done it."

One method to attenuate murdercide, then, is to target dangerous groups, such as al-Qaeda, that influence individuals. Another method, says Princeton University economist Alan B. Krueger, is to increase the civil liberties of the countries that breed terrorist groups. In an analysis of State Department data on terrorism, Krueger discovered that "countries like Saudi Arabia and Bahrain, which have spawned relatively many terrorists, are economically well off yet lacking in civil liberties. Poor countries with a tradition of protecting civil liberties are unlikely to spawn terrorists. Evidently, the freedom to assemble and protest peacefully without interference from the government goes a long way to providing an alternative to terrorism."

Let freedom ring.

51

As Luck Would Have It

Are some people really luckier than others, or is it all in their heads? Both

Amyotrophic lateral sclerosis (ALS) is a neuromuscular disease that attacks motor neurons until muscle weakness, atrophy, and paralysis lead inexorably to death. Victims of this monstrous malady could be forgiven for thinking themselves unlucky.

How, then, can we explain the attitude of the disease's namesake, baseball great Lou Gehrig? He told a sellout crowd at Yankee Stadium: "For the past two weeks you have been reading about the bad break I got. Yet today I consider myself the luckiest man on the face of this earth." The Iron Horse went on to recount his many blessings and fortunes, a list twice punctuated with "I'm lucky" and "that's something." Choking back the emotional gravitas, Gehrig concluded that "I may have been given a bad break, but I have an awful lot to live for."

The physical meltdown caused by ALS was documented on ABC's *Nightline*, as Brandeis University sociologist Morrie Schwartz turned his plight into an opportunity to teach one final course on the lessons of life. "Through my dying I'm teaching people how to live," he told his old student Mitch Albom, whose book *Tuesdays with Morrie* records the life wisdom of a dying man. "I can't go shopping, I can't take out the garbage, I can't take care of the bank accounts," Schwartz admitted, "but I can take care and look at what I think is important in life, and I have both the leisure and the time and the impulse to do that."

Clearly luck is a state of mind. Is it more than that? To explore this question scientifically, the experimental psychologist Richard Wiseman created a "luck lab" at the University of Herfordshire in Britain. Wiseman began by testing whether lucky people are actually luckier in winning the lottery. He recruited seven hundred people who had already purchased lottery tickets to complete his luck questionnaire, which is a self-report scale that measures whether people consider themselves to be lucky or unlucky. Although lucky people were twice as confident as unlucky people that they would win the lottery, there was no difference in winnings. Wiseman then administered to another group of subjects a standardized IQ test and found no difference in intelligence between those who considered themselves to be lucky or unlucky.

Wiseman then gave subjects a standardized "life satisfaction" scale that asks people to rank themselves on how satisfied they are with their family life, personal life, financial situation, health, and career. The results were striking. "Lucky people are far more satisfied with all areas of their lives than unlucky or neutral people," Wiseman reveals in his charming and insightful book *The Luck Factor* (Miramax Books, 2003). Does this satisfied state of mind translate into actual life outcomes that someone might call "lucky"? It does. Here's how.

Wiseman gave subjects the "Big Five" personality scale, which measures *agreeableness*, *conscientiousness*, *extroversion*, *neuroticism*, and *openness*. Although there were no differences on *agreeableness* and *conscientiousness*, Wiseman found statistically significant differences between lucky and unlucky people on *extroversion*, *neuroticism*, and *openness*.

Lucky people score significantly higher than unlucky people on *extroversion*. "There are three ways in which lucky people's extroversion significantly increases the likelihood of their having a lucky chance encounter," Wiseman explains, "meeting a large number of people, being a 'social magnet,' and keeping in contact with people." Lucky people, for example, smile twice as much and engage in more eye contact than unlucky people, which leads to more social encounters, which generates more opportunities.

The personality dimension of *neuroticism* measures how anxious or relaxed someone is, and Wiseman found that lucky people were half as anxious as unlucky people; that is, "because lucky people tend to be more relaxed than most, they are more likely to notice chance opportunities, even when they are not expecting them." In one experiment, Wiseman had subjects count the number of photographs in a newspaper. Lucky people were more likely than unlucky people to notice on page two a half-page ad with this message in large, bold type: STOP COUNTING—THERE ARE 43 PHOTOGRAPHS IN THIS NEWSPAPER.

Wiseman discovered that lucky people also score significantly higher in *openness* than unlucky people. "Lucky people are open to new experiences in their lives. They don't tend to be bound by convention and they like the notion of unpredictability." As such, lucky people travel more, encounter novel prospects, and welcome unique opportunities.

Expectation also plays a role in luck. Lucky people expect good things to happen, and when they do, they embrace them. But even when misfortune falls, lucky people turn bad luck into good fortune. Consider the example set by the longest ALS sufferer in history, Stephen Hawking, who writes, "I was fortunate that my scientific reputation increased, at the same time that my disability got worse. This meant that people were prepared to offer me a sequence of positions in which I only had to do research, without having to lecture." That led to his cognitive style of thinking through problems visually and geometrically, instead of computationally on a chalkboard, which was no longer available to him. "I was lucky to have chosen to work in theoretical physics, because that was one of the few areas in which my condition would not be a serious handicap." Confined to an electric wheelchair and unable to move, Hawking capitalized on his ill fortune by using it as a chance to revolutionize science and transform the universe, which he did.

That's something.

52

SHAM Scam

The Self-Help and Actualization Movement is an $8.5-billion-a-year business. Does it work?

According to self-help guru Tony Robbins, walking across thousand-degree red-hot coals barefoot "is an experience in belief. It teaches people in the most visceral sense that they can change, they can grow, they can stretch themselves, they can do things they never thought possible."

I've done three firewalks myself, without chanting "cool moss" (as Robbins has his clients do) or thinking positive thoughts. I didn't get burned. Why? Because wood is a poor conductor of heat, particularly through the dead, calloused skin on the bottom of your feet, and especially if you scoot across the bed of coals as quickly as firewalkers are wont to do. Think of a cake in a four-hundred-degree oven—you can touch the poor-conducting cake without getting burned, but not the metal cake pan. Physics explains the "how" of firewalking. To understand the "why" we must turn to psychology.

In 1980 I attended a bicycle industry trade convention whose keynote speaker was Mark Victor Hansen, now well known as the coauthor of the wildly popular *Chicken Soup for the Soul* book series that includes the *Teenage Soul*, *Prisoner's Soul*, and *Christian's Soul* (but no *Skeptic's Soul*). I was surprised that Hansen didn't require a speaker's fee until I saw what happened after his talk: people were lined up out the door to purchase his motivation tapes. I was one of them. I listened to those tapes over and over during training rides in preparation for bicycle races.

The "over and over" part is the key to understanding the "why" of what investigative journalist Steve Salerno calls the Self-Help and Actualization Movement (SHAM). In his book *Sham: How the Self-Help Movement Made America Helpless* (Crown, 2006), he explains how the talks and tapes offer a momentary boost of inspiration that fades after a few weeks, turning buyers into repeat customers. When Salerno was a self-help-book editor for Rodale Press (whose motto was "to show people how they can use the power of their bodies and minds to make their lives better"), extensive market surveys revealed that "the most likely customer for a book on any given topic was someone who had bought a similar book within the preceding eighteen months." The irony of "the Eighteen-Month Rule" for this genre, says Salerno, is this: "If what we sold worked, one would expect lives to improve. One would not expect people to need further help from us—at least not in that same problem area, and certainly not time and time again."

Surrounding SHAM is a bulletproof shield: if your life does not improve it is your fault—your thoughts were not positive enough. The solution? More of the same, or at least the same message repackaged into new products, as in the multiple permutations of John Gray's *Men Are from Mars, Women Are from Venus, Mars and Venus Together Forever, Mars and Venus in the Bedroom, The Mars and Venus Diet and Exercise Solution*—not to mention the *Mars and Venus* board game, musical, and Club Med getaway.

SHAM takes advantage of a clever marketing dualism of victimization and empowerment. Like a religion that defines people as inherently sinful so that they require forgiveness (provided exclusively by that religion), SHAM gurus insist that we are all victims of our demonic "inner children" produced by traumatic pasts that create negative "tapes" that replay over and over in our minds. Redemption comes through empowering yourself with new "life scripts," supplied by the seminarists themselves, for prices that range from $500 one-day workshops to Robbins's $6,995 "Date with Destiny" seminar.

Do these programs work? It depends on how one defines "work," and herein lies the problem with trying to answer this most fundamental question about a movement that has grown from the benign counsel of Dale Carnegie's 1937 *How to Win Friends and Influence People* to today's $8.5-billion-a-year industry. Although motivation seminarists publish countless glowing testimonials in their promotional literature, there is no scientific evidence showing that self-help programs work, and some reasons to think that they may do more harm than good. According to Salerno, none of the countless SHAM programs—from firewalking to twelve-stepping—work better than doing something else, or even doing nothing. The law of large numbers means that given the millions of people who have tried SHAM programs, it is inevitable that some will improve. But millions more people have never tried any self-help schemes, and some of them improve as well. Is there any difference between these two populations? Yes. One of them is lighter in the pocketbook.

As with alternative medicine nostrums, the body naturally heals itself, and whatever the patient was doing at the time gets the credit. Patient, heal thyself—the true meaning of self-help.

53

The Political Brain

A recent brain-imaging study shows that our political predilections are products of unconscious confirmation bias

> The human understanding when it has once adopted an opinion . . . draws all things else to support and agree with it. And though there be a greater number and weight of instances to be found on the other side, yet these it either neglects and despises . . . in order that by this great and pernicious predetermination the authority of its former conclusions may remain inviolate.
>
> —Francis Bacon, *Novum Organum*, 1620

Pace Will Rogers, I am not a member of any organized political party. I am a libertarian. As a fiscal conservative and social liberal, I never met a Republican or Democrat in whom I could not find something to like. I have close friends in both camps in which I have observed the following: no matter the issue under discussion, both sides are equally convinced that the evidence overwhelmingly supports their position.

This surety is the confirmation bias, where we seek and find confirmatory evidence in support of already existing beliefs and ignore or reinterpret disconfirmatory evidence. According to Tufts University psychologist Raymond Nickerson, in a comprehensive literature review ("Confirmation Bias: A Ubiquitous Phenomenon in Many Guises," *Review of General Psychology* 2, no. 2 [1998]: 175–220), the confirmation bias "appears to be sufficiently strong and pervasive that one is led to wonder whether the bias, by itself, might account for a significant fraction of the disputes, altercations, and misunderstandings that occur among individuals, groups, and nations."

Now a functional magnetic resonance imaging (fMRI) study shows where in the brain the confirmation bias occurs, and how it is unconscious and driven by emotions. The study was conducted at Emory University under the direction of psychologist Drew Westen, and the results were presented at the January 28, 2006, Annual Conference of the Society for Personality and Social Psychology.

During the run-up to the 2004 presidential election, while undergoing an fMRI bran scan, thirty men—half self-described "strong" Republicans and half "strong" Democrats—were tasked with assessing statements by both George W. Bush and John Kerry in which the candidates clearly contradicted themselves. Not surprisingly, in their assessments Republican subjects were as critical of Kerry as Democratic subjects were of Bush, yet both let their own preferred candidate off the evaluative hook.

The neuroimaging results, however, revealed that the part of the brain most associated with reasoning—the dorsolateral prefrontal cortex—was quiescent. Most active were the orbital frontal cortex, which is involved in the processing of emotions; the anterior cingulated, which is associated with conflict resolution; the posterior cingulated, which is concerned with making judgments about moral accountability; and—once subjects had arrived at a conclusion that made them emotionally comfortable—the ventral striatum, which is related to reward and pleasure.

"We did not see any increased activation of the parts of the brain normally engaged during reasoning," Westen explained. "What we saw instead was a network of emotion circuits lighting up, including circuits hypothesized to be involved in regulating emotion, and circuits known to be involved in resolving conflicts." Interestingly, neural circuits engaged in rewarding selective behaviors were activated. "Essentially, it appears as if partisans twirl the cognitive kaleidoscope until they get the conclusions they want, and then they get massively reinforced for it, with the elimination of negative emotional states and activation of positive ones."

These are the neural correlates of the confirmation bias, and the implications reach far beyond politics. A judge or jury assessing evidence against a defendant, a CEO evaluating information about a company, or a scientist weighing data in favor of a theory will undergo the same cognitive process. What can we do about it?

In science we have built-in self-correcting machinery. Strict double-blind controls are required in experiments, in which neither the subjects nor the experimenters know the experimental conditions during the data collection phase. Results are vetted at professional conferences and in peer-reviewed journals. Research must be replicated in other labs unaffiliated with the original researcher. Disconfirmatory evidence, as well as contradictory interpretations of the data, must be included in the paper. Colleagues are rewarded for being skeptical. Extraordinary claims require extraordinary evidence. As Westen notes, however, "Even with these safeguards in place, scientists are prone to confirmatory biases, particularly when reviewers and authors share similar beliefs, and studies have shown that they will judge the same methods as satisfactory or unsatisfactory depending on whether the results matched their prior beliefs."

We need similar controls for the confirmation bias in the law, business, and politics. Judges and lawyers should call each other out on the practice of mining data selectively to bolster an argument and warn juries about the confirmation bias. CEOs should assess skeptically the enthusiastic recommendations of their VPs and demand to see contradictory evidence and alternative evaluations of the same plan. Politicians need a stronger peer-review system that goes beyond the churlish opprobrium of the campaign trail, and I would love to see a political debate in which the candidates were required to make the opposite case.

Skepticism is the antidote for the confirmation bias.

54

Folk Science

Why our intuitions about how the world works are often wrong

Thirteen years after the legendary confrontation over the theory of evolution between Bishop Samuel Wilberforce ("Soapy Sam") and Thomas Henry Huxley ("Darwin's bulldog"), Wilberforce died in an equestrian fall in 1873. Huxley quipped to the physicist John Tyndall, "For once, reality and his brain came into contact and the result was fatal."

When it comes to such basic forces as gravity and such fundamental phenomena as falling, our intuitive sense of how the physical world works—our folk physics—is reasonably sound. Thus we appreciate Huxley's wry comment and note that even children get the humor of cartoon physics where, for example, when a character runs off a cliff he does not fall until he realizes that he has left terra firma (also known as "coyotes interruptus," in honor of Wile E. Coyote, who frequently fell to his doom in this manner while chasing his road runner nemesis).

But much of physics is counterintuitive, as is the case in many other disciplines as well, and before the rise of modern science we had only our folk intuitions to guide us. Folk astronomy, for example, told us that the world is flat, celestial bodies revolve around the Earth, and the planets are wandering gods who determine our future. Folk biology intuited an élan vital flowing through all living things, which in their functional design were believed to have been created ex nihilo by an intelligent designer. Folk psychology compelled us to search for the homunculus in the brain—a ghost in the machine—a mind somehow disconnected from the brain. Folk economics caused us to disdain excessive wealth, label usury as a sin, and mistrust the invisible hand of the market.

The reason why folk science so often gets it wrong is that we evolved in an environment radically different from the one in which we live. Our senses are geared for perceiving objects of middling size—between, say, ants and mountains—not bacteria, molecules, and atoms on one end of the scale, and stars and galaxies on the other end. We live a scant three score and ten years, far too short a time to witness evolution, continental drift, or long-term environmental changes.

Causal inference in folk science is equally untrustworthy. We correctly surmise designed objects such as stone tools to be products of an intelligent designer, and thus naturally assume that all functional objects, such as eyes, must have been similarly intelligently designed. Lacking a cogent theory of how neural activity gives rise to consciousness, we imagine mental spirits floating within our heads. We lived in small bands of roaming hunter-gatherers who accumulated little wealth and had no experience of free markets and economic growth.

More generally, folk science leads us to trust anecdotes as data, such as illnesses being cured by assorted nostrums based solely on single-case examples. Equally powerful are anecdotes involving preternatural beings, compelling us to make causal inferences linking these nonmaterial entities to all manner of material events, illness being the most personal. Because people often recover from illness naturally, whatever was done just before recovery receives the credit, prayer being the most common.

In this latter case we have a recent real science analysis of this ancient folk science supposition. The April 2006 issue of *American Heart Journal* published a comprehensive study directed by Harvard University Medical School cardiologist Herbert Benson on the effects of intercessory prayer on the health and recovery of patients undergoing coronary bypass surgery. The 1,802 patients were divided into three groups, two of which were prayed for by members of three religious congregations. Prayers began the night before the surgery and continued daily for two weeks after. The prayers were allowed to pray in their own manner, but they were instructed to ask "for a successful surgery with a quick, healthy recovery and no complications." Half the prayer-recipient patients were told that they were being prayed for while the other half were told that they might or might not receive prayers. Results showed no statistically significant differences between any of the groups. Case closed.

Of course, people will continue praying for their ill loved ones, and by chance some of them will recover, and our folk science brains will find meaning in these random patterns. But to discriminate true causal inferences from false, real science trumps folk science.

55

Free to Choose

The neuroscience of choice exposes the power of ideas

Have you ever watched a white rat choose between an 8 percent and a 32 percent sucrose solution by pressing two different bars on variable-interval schedules of reinforcement? No? Lucky you. I devoted two years of what would otherwise have been a misspent youth to running choice experiments with rats in Skinner boxes at California State University–Fullerton under the direction of Douglas J. Navarick for a master's thesis in 1978, "Choice in Rats as a Function of Reinforcer Intensity and Quality." Boys gone wild!

Since then, behaviorists' black box has been penetrated by neuroscientists, most recently by Read Montague from the Institute of Advanced Study in *Why Choose This Book?* (Dutton, 2006). Montague argues that our brains evolved computational programs to evaluate choices in terms of their value and efficiency: "Those that accurately estimate the costs and the long-term benefits of choice will be more efficient than those that don't—and in the long term these are the winners."

Life, like the economy, is about the efficient allocation of limited resources that have alternative uses (to paraphrase the economist Thomas Sowell). It all boils down to energy efficiency. To a predator, says Montague, prey are batteries of energy. "This doctrine mandates that evolution discover efficient computational systems that know how to capture, process, store, and reuse energy efficiently." Those that do, pass on their genetic programs for efficient computational neural processing to make efficient choices. Over the course of millions of years, says Montague, our brain has evolved to be so efficient that it consumes about a fifth of the energy of an average lightbulb, costing about a nickel a day to run.

Computational programs are designed by evolution to learn how to solve certain tasks. Rats, for example, inherit programs that are especially good at learning mazes and pressing bars because they evolved to forage in dark and spatially complex environments. There are no blank slates for mice or men. "Despite their differences," Montague explains, "all goals have one thing in common: They can all be used by our brains to direct decisions that lead to the satisfaction of the goal."

Unfortunately, these evolved computational programs can be hijacked. Addictive drugs, for example, rewire the brain's dopamine system—normally used to reward choices that are good for the organism, such as food, family, and friends—to reward choosing the next high instead. Ideas do something similar, in that they take over the role of reward signals that feed into the dopamine neurons. This includes *bad* ideas, such as the Heaven's Gate cult members who chose suicide to join the mother ship they believed was awaiting them near the comet Hale-Bopp. The brains of suicide bombers have been similarly commandeered by religious and political bad ideas.

In *The Science of Good and Evil* (Times Books, 2004) I argued that we evolved moral emotions that operate similar to other emotions, such as hunger and sexual appetite. Thinking of these emotions as proxies for highly efficient computational programs deepens our understanding of the process. When we need energy we do not compute the relative caloric values of our food choices; we just feel hungry for certain food types, eat them, and are rewarded with a sense of satisfaction. Likewise, in choosing a sexual partner, the brain employs a computational program to make you feel attracted to people with good genes, as indicated by such proxies as a symmetrical face and body, clear complexion, and a 0.7 waist-to-hip ratio in women and an inverted pyramid build in men. Similarly, in making moral choices about whether to be altruistic or selfish, we feel guilt or pride for having done the wrong or the right thing, but the moral calculation of what is best for the individual and the social group was made by our Paleolithic ancestors. Emotions such as hunger, lust, and pride are stand-ins for these computations.

How can we utilize this theory of choice to our advantage? Montague employed fMRI brain scans to discover that certain brands, such as Coke, "change dopamine delivery to various brain regions through their effect on reward prediction circuitry." The Coke brand has a "flavor" in the ventromedial prefrontal cortex, a region essential for decision making. Just as Coke is a proxy for flavor, hunger a proxy for caloric need, lust a proxy for reproductive necessity, and guilt and joy proxies for immoral and moral behavior, so too can we market moral brands to rewire brains to value and choose good ideas.

In honor of the late economist Milton Friedman, author of the radical book *Free to Choose*, I propose that we begin by marketing this brand—the Principle of Freedom: *All people are free to think, believe, and act as they choose, as long as they do not infringe on the equal freedom of others.*

56

Bush's Mistake and Kennedy's Error

Self-deception proves itself to be more powerful than deception

The war in Iraq is now four years old. At a cost of $200 million a day, $73 billion a year, and nearly $300 billion since it began, plus more than 3,000 American lives, that's a substantial investment. No wonder most members of Congress from both parties, along with President Bush, believe that we've got to "stay the course" and not just "cut and run." As Bush explained in a Fourth of July 2006 speech at Fort Bragg, North Carolina, "I'm not going to allow the sacrifice of 2,527 troops who have died in Iraq to be in vain by pulling out before the job is done." (This essay was originally published in May 2007. The above figures are all considerably higher.)

We all make similarly irrational arguments about decisions in our lives: we hang on to losing stocks, unprofitable investments, failing businesses, and unsuccessful relationships. But if we were rational we would just compute the odds of succeeding from this point forward and then decide if the investment warrants the potential payoff. But we are not rational—not in love, or war, or business—and this particular irrationality is what economists call the "sunk-cost fallacy."

The psychology beneath this and other cognitive fallacies is brilliantly illuminated in *Mistakes Were Made (but Not by Me)*, by psychologists

Carol Tavris and Elliot Aronson (Harcourt, 2007). They focus on self-justification, which "allows people to convince themselves that what they did was the best thing they could have done." The passive past tense of the all-telling phrase—mistakes were made—shows the rationalization process at work. "Mistakes were quite possibly made by the administrations in which I served," confessed Henry Kissinger about Vietnam, Cambodia, and South America. "If, in hindsight, we also discover that mistakes may have been made . . . I am deeply sorry," admitted Cardinal Edward Egan of New York about the Catholic Church's failure to deal with priestly pedophiles.

The engine driving self-justification is cognitive dissonance, "a state of tension that occurs whenever a person holds two cognitions (ideas, attitudes, beliefs, opinions) that are psychologically inconsistent," Tavris and Aronson explain. "Dissonance produces mental discomfort, ranging from minor pangs to deep anguish; people don't rest easy until they find a way to reduce it." It is in that process of reducing dissonance that the self-justification accelerator is throttled up.

Leon Festinger first discovered cognitive dissonance during his investigation of a UFO cult that firmly believed a mother ship from planet Clarion would arrive on December 20, 1954, just in time to whisk them away to safety when the world ended the next day. When doomsday came and went without incident (or mother ship), Festinger's prediction was fulfilled: those members who made the biggest commitment to the cult by quitting their jobs, leaving their spouses, and giving away their possessions would be the least likely to admit their error. In fact, after the non-event these true believers announced that their unwavering faith had saved the world! Dissonance resolved.

Wrongly convicting people and sentencing them to death is a supreme source of cognitive dissonance. Since 1992, the Innocence Project has exonerated 188 people from death row. "If we reviewed prison sentences with the same level of care that we devote to death sentences," says the University of Michigan law professor Samuel R. Gross, "there would have been *over 28,500 non-death-row exonerations in the past 15 years*, rather than the 255 that have in fact occurred." What is the self-justification for reducing this form of dissonance? "You get in the system and you become very cynical," explains Northwestern University law professor Rob

Warden. "People are lying to you all over the place. Then you develop a theory of the crime, and it leads to what we call tunnel vision. Years later overwhelming evidence comes out that the guy was innocent. And you're sitting there thinking, 'Wait a minute. Either this overwhelming evidence is wrong or I was wrong—and I couldn't have been wrong because I'm a good guy.' That's a psychological phenomenon I have seen over and over."

What happens in those rare instances when someone says "I was wrong"? Surprisingly, forgiveness is granted and respect is elevated. Imagine what would happen if President George W. Bush had publicly said the following:

> This administration intends to be candid about its errors. For as a wise man once said, "An error does not become a mistake until you refuse to correct it." We intend to accept full responsibility for our errors. We're not going to have any search for scapegoats . . . the final responsibility of any failure is mine, and mine alone.

Bush's popularity would have skyrocketed, and respect for his ability as a thoughtful leader willing to change his mind in the teeth of new evidence would have soared. We know, because that is precisely what happened to President John F. Kennedy after the botched Bay of Pigs invasion of Cuba, when he publicly spoke the above words.

Update: By the end of the wars in Iraq and Afghanistan a total of 5,281 American soldiers and 1,432 other personnel died, for a total of 6,713. According to a 2013 report conducted by Harvard University's Kennedy School the combined costs of both wars are $4 trillion to $6 trillion. Of the 1.6 million troops dispatched in the decade-long conflicts, more than half received medical treatment and will receive benefits for the rest of their lives, which the report estimated will add $836 billion to the bill. At such a price the sunk-cost bias will remain strong for decades to come.

VIII

HUMAN NATURE

57

The Erotic-Fierce People

The latest skirmish in the "anthropology wars" reveals a fundamental flaw in how science is understood and communicated

Another battle has broken out in the century-long "anthropology wars" over the true nature of human nature. Journalist Patrick Tierney, in a widely reviewed book dramatically titled *Darkness in El Dorado: How Scientists and Journalists Devastated the Amazon*, purportedly reveals "the hypocrisy, distortions, and humanitarian crimes committed in the name of research, and reveals how the Yanomamö internecine warfare was, in fact, triggered by the repeated visits of outsiders who went looking for a 'fierce' people whose existence lay primarily in the imagination of the West."

Tierney's bête noire is Napoleon Chagnon, whose ethnography *Yanomamö: The Fierce People* has been the best-selling anthropological book of all time. Befitting his thesis, Tierney spares no ink in painting a picture of Chagnon as a fierce anthropologist who sees in the Yanomamö nothing more than a reflection of himself. Chagnon's sociobiological theories of the most violent and aggressive males winning the most copulations and thus passing on their genes for "fierceness," says Tierney, is nothing more than a window into Chagnon's own libidinous impulses.

Are the Yanomamö the "fierce people"? Or are they the "erotic people," as described by the French anthropologist Jacques Lizot, another of

Tierney's targets? The problem lies in the phrasing of the question. Humans are not so easily pigeonholed into such clear-cut categories as "fierce" or "erotic." Clearly we are both (and a lot more), the nature and intensity of behavioral expression being dependent on a host of biological, social, and historical variables. Chagnon understands this. Tierney does not. Thus *Darkness in El Dorado* fails not just because he didn't get the story straight (although there are countless factual errors and intentional distortions in the book) but also because of a fundamental misunderstanding of how science works and the difference between isolated anecdotes (upon which his book is based) and statistical trends (upon which Chagnon's book depends).

To be sure, Tierney is a good storyteller, but this is what makes his attack on science so invidious. Since humans are storytelling animals, we are more readily convinced by dramatic anecdotes than we are by dry data. I must admit I was enraged by many of his stories . . . until I checked the sources myself. For example, I read Chagnon's classic ethnography and discovered that the fourth edition carries no subtitle. Had Chagnon determined that the Yanomamö were not the "fierce people" after all? No. He realized that too many people were unable to move past the moniker to grasp the complex and subtle variations in all human populations, and they "might get the impression that being 'fierce' is incompatible with having other sentiments or personal characteristics like compassion, fairness, valor, etc." As he notes in his opening chapter, the Yanomamö "are simultaneously peacemakers and valiant warriors." Like all people, the Yanomamö have a deep repertoire of responses for varying social interactions and differing contexts.

Tierney accuses Chagnon of using the Yanomamö to support a sociobiological model of an aggressive human nature. Yet the primary sources in question show that Chagnon's deductions from the data are not so crude. Even on the final page of his chapter on Yanomamö warfare, Chagnon inquires about "the likelihood that people, throughout history, have based their political relationships with other groups on predatory versus religious or altruistic strategies and the cost-benefit dimensions of what the response should be if they do one or the other." He concludes, "We have the evolved capacity to adopt either strategy." These are hardly the words of a hidebound ideologue bent on indicting the human species.

After interviewing all the major players in this drama, and plowing through much of the anthropological literature, my conclusion is that Chagnon's view of the Yanomamö is basically supported by the available evidence. His data and interpretations are corroborated by many other anthropologists. Even at their "fiercest," however, the Yanomamö are not so different from many other nonstate peoples around the globe (recall Captain Bligh's and Captain Cook's numerous violent encounters with Polynesians). And, judging by the latest archaeological research, Yanomamö violence is certainly no more extreme than that of our Paleolithic ancestors who appear to have brutally butchered one another with reckless abandon. If the past five thousand years of recorded history are any measure of a species' "fierceness," the Yanomamö have nothing on Western "civilization," whose record of killing includes hundreds of millions murdered in organized violence.

Homo sapiens in general, like the Yanomamö in particular, are erotic-fierce people, making love and war far too frequently for our own good as both overpopulation and war threaten our very existence. Fortunately we now have the scientific tools to not only illuminate our true natures, but also to help us navigate the treacherous shoals of surviving the transition from a state society to whatever comes next.

Update: Over the years a number of independent investigations were made into the charges against Chagnon in Tierney's book. The University of Michigan, where he was a professor, for example, found the claims baseless, and the historian of science Alice Dreger concluded that Tierney's accusations were false and that the American Anthropological Association was complicit and irresponsible in siding with Tierney before the facts were in, and for not protecting "scholars from baseless and sensationalistic charges." http://bit.ly/1q04vbZ

58

The Ignoble Savage

Science reveals humanity's heart of darkness

In 1670, the British poet John Dryden penned this expression of humans in a state of nature: *I am as free as Nature first made man / When wild in woods the noble savage ran.* Eighty-five years later, in 1755, the French philosopher Jean-Jacques Rousseau canonized the noble savage into Western culture by proclaiming, "nothing can be more gentle than him in his primitive state, when placed by nature at an equal distance from the stupidity of brutes and the pernicious good sense of civilized man."

From the Disneyfication of Pocahontas to Kevin Costner's eco-pacifist Native Americans in *Dances with Wolves*, and from postmodern accusations of corrupting modernity to modern anthropological theories that indigenous people's wars are just ritualized games, the noble savage remains one of the last epic creation myths of our time.

Science reveals a rather different picture of humanity in its natural state. In a 1996 study, University of Michigan ecologist Bobbi Low analyzed 186 hunter-gatherer societies and discovered that their relatively low environmental impact is the result of low population density, inefficient technology, and the lack of profitable markets, not from conscious efforts at conservation. Anthropologist Shepard I. Krech, in *The Ecological Indian*, shows that in a number of Native American communities large-scale irrigation practices led to the salinization and exhaustion of river valleys, ultimately triggering the collapse of their societies.

Even the reverence for big game animals we have been told was held by Native Americans is a myth—most believed that common game animals such as elk, deer, caribou, beaver, and especially buffalo would be physically reincarnated by the gods. Given the opportunity to hunt big game animals to extinction, they did. The evidence is now overwhelming that woolly mammoths, giant mastodons, ground sloths, one-ton armadillo-like glyptodonts, bear-sized beavers, and beefy sabertooth cats, not to mention American lions, cheetahs, camels, horses, and many other large mammals, all went extinct at the same time that Native Americans first populated the continent in the mass migration from Asia. The best theory to date as to the cause of this mass extinction is overhunting.

Ignoble savages were nasty to each other as well as their environments. Violence, aggression, and war are part of the behavioral repertoire of the great apes and humans. The propensity for a band of young and aggressive males (whether human or chimpanzee) to fan out into neighboring environments on a seek-and-destroy mission to gain resources and females is clearly there in the species, the genus, and the family.

Surveying primitive and civilized societies, University of Illinois anthropologist Lawrence H. Keeley, in *War Before Civilization*, demonstrates that prehistoric war was, relative to population densities and fighting technologies, at least as frequent (measured in years at war versus years at peace), as deadly (measured by percentage of conflict deaths within a population), and as ruthless (measured by the killing and maiming of noncombatants, women, and children) as modern war. One pre-Columbian mass grave in South Dakota, for example, yielded the remains of five hundred scalped and mutilated men, women, and children.

In *Constant Battles*, a recent and exceptionally insightful study of this problem by Steven A. LeBlanc, the Harvard archaeologist quips, "anthropologists have searched for peaceful societies much like Diogenes looked for an honest man." Consider the evidence from a ten-thousand-year-old Paleolithic site along the Nile River: "The graveyard held the remains of 59 people, at least 24 of whom showed direct evidence of violent death, including stone points from arrows or spears within the body cavity, and many contained several points. There were six multiple burials, and almost all those individuals had points in them, indicating that the people in each mass grave were killed in a single event and then buried together."

LeBlanc's survey of our not-so-noble past reveals that even cannibalism, long thought to be a form of primitive urban legend to be debunked (noble savages would never eat each other, would they?), has now been supported by powerful physical evidence that includes broken and burned bones; cut marks on the bone; bones broken open lengthwise to get at the marrow; and, inside cooking jars, bones broken to a length so that they would fit. Such evidence for prehistoric cannibalism has been uncovered in Mexico, Fiji, and in Spain and other parts of Europe. The final (and gruesome) proof came with the discovery of the human muscle protein myoglobin in the fossilized human feces of a prehistoric Anasazi pueblo Indian.

Savage, yes. Noble, no.

The Roman statesman Cicero remarked, "Although physicians frequently know their patients will die of a given disease, they never tell them so. To warn of an evil is justified only if, along with the warning, there is a way of escape." As we shall see in the next essay, there is an escape from our disease.

59

The Domesticated Savage

Science reveals a way to rise above our natures

Nature, Mr. Alnutt, is what we were put in this world to rise above.

—Katharine Hepburn to Humphrey Bogart in *The African Queen*

The UCLA evolutionary biologist Jared Diamond once classified humans as the "third chimpanzee" (the second being the bonobo). Genetically we are very similar, and when it comes to high levels of aggression between members of two different groups, as we saw in the previous chapter, "The Ignoble Savage," we also resemble chimpanzees. While humans have a brutal history, there's hope that the pessimists who forecast our eventual demise are wrong; recent evidence indicates that, like bonobos, we may be evolving in a more peaceful direction.

In his 1859 book *The Origin of Species*, Charles Darwin made an analogy between artificial selection by breeders and natural selection by nature. Consider this analogue. It has been observed that one of the most striking features in artificially selecting for docility among wild animals is that, along with far less aggression, you also get a suite of other changes, including and especially a reduction in skull, jaw, and teeth size. In genetics, this is called pleiotropy. Selecting for one trait (nonaggression) may generate other unintended changes (reduced skull, jaw, and teeth size).

The most famous study on selective breeding for passivity was begun in 1959 by the Russian geneticist Dmitri Belyaev, at the Institute of Cytology and Genetics in Siberia (and continues today under the direction of

Lyudmila N. Trut), in which silver foxes were bred for friendliness toward humans (defined by a graduating series of criteria, from the animal allowing itself to be approached, to being hand fed, to being petted, to proactively seeking human contact). In only thirty-five generations (remarkably short on an evolutionary time scale) the researchers were able to produce tail-wagging, hand-licking, peaceful foxes. What they also fashioned were foxes with smaller skulls, jaws, and teeth than their wild ancestors. (Similar changes can be seen in comparing domesticated dogs to their ancestral wolves.)

What is going on here? The Russian scientists believe that in selecting for docility, they inadvertently selected for paedomorphism—the retention of juvenile features into adulthood—such as curly tails and floppy ears (found in wild pups but not in wild adults), a delayed onset of the fear response to unknown stimuli, and lower levels of aggression. The selection process led to a significant decrease in levels of stress-related hormones such as corticosteroids, which are produced by the adrenal glands during the flight-or-fight response, as well as a significant increase in levels of serotonin, thought to play a leading role in the inhibition of aggression. Curiously, in selecting only for tameness, the Russian scientists were also able to accomplish what no breeder had achieved before—increase the length of the breeding season.

What does this analogy tell us? It turns out that while humans are like chimpanzees when it comes to between-group aggression, when we compare levels of aggression among members of the same social group, it turns out that we are much more like peaceful, highly sexual bonobos. Harvard anthropologist Richard Wrangham proffers a plausible theory for this contrast: as a result of selection pressures for greater within-group peacefulness and sexuality, humans and bonobos have gone down a different behavioral evolutionary path than chimps. That difference may be witnessed in morphology. Bonobos were once called the "pygmy chimpanzee" because, compared to chimps, their skull, jaws, and teeth are much reduced in size.

Wrangham suggests that over the past twenty thousand years, as humans became more sedentary and their populations grew, there was selection pressure for less within-group aggression, and this effect can be seen in such paedomorphic features as smaller skulls, jaws, and teeth (compared to our immediate hominid ancestors), as well as our year-round breeding season and prodigious sexuality. (Emory University psychologist Frans B. de Waal, in his extensive study of primate social behavior, has shown how bonobos in particular use sexual contact as an important form of conflict resolution and social bonding.) Wrangham also shows how "Area 13" in the limbic frontal cortex of humans, believed to mediate aggression, more closely resembles in size the equivalent area in bonobo brains than it does in that same area in chimpanzee brains.

A plausible evolutionary hypothesis suggests itself. Limited resources led to the selection for within-group cooperation and between-group competition in humans, resulting in within-group amity and between-group enmity. This evolutionary scenario bodes well for our species if we can continue to expand the circle of whom we consider to be members of our in-group. We have a long way to go still, and recent tribal conflicts and religious wars are not encouraging, but if we adopt the evolutionist's stock in trade—deep time—we can see that over the past millennium there has been a long-term trend toward including more people (e.g., women and minorities) into the in-group deserving of human rights.

60

A Bounty of Science

A new book reexamines the mutiny on the Bounty,
but science offers a deeper account of its cause

The most common explanation for the mutiny on the *Bounty* pits a humane Fletcher Christian against an oppressive William Bligh. In her 2003 revisionist book *The Bounty*, Caroline Alexander recasts Bligh as hero and Christian as coward. After four hundred pages of gripping narrative, Alexander wonders "what caused the mutiny," hints that it might have had something to do with "the seductions of Tahiti" and "Bligh's harsh tongue," but then concludes that it was "a night of drinking and a proud man's pride, a low moment on one gray dawn, a momentary and fatal slip in a gentleman's code of discipline."

A skeptic's explanation may seem less romantic, but it is ultimately more intellectually satisfying because it is extrapolated from scientific evidence and reasoning. There are, in fact, two levels of causality we can examine: (1) *proximate* (immediate historical events) and (2) *ultimate* (deeper evolutionary motives). A quantitative analysis of every lash British sailors received from 1765 through 1793 while serving on fifteen naval vessels that sailed into the Pacific shows that Bligh was not overly abusive when compared with his contemporaries. The Australian historian Greg Dening, in his book *Mr. Bligh's Bad Language*, computed the average percentage of sailors flogged at 21.5. Bligh's average was 19 percent, lower than James Cook's 20, 26, and 37 percent, respectively, on his three voyages, and less than half of George Vancouver's 45 percent. Vancouver averaged 21 lashes per sailor, compared with the overall mean of 5 and Bligh's mean of only 1.5.

If unusually harsh punishment was not the deeper cause of the mutiny, then what was? Although Bligh preceded Darwin by a century, he comes closest, in my opinion, to capturing the ultimate cause: "I can only conjecture that they have Idealy assured themselves of a more happy life among the Otaheitians than they could possibly have in England, which joined to some Female connections has most likely been the leading cause of the whole business."

Indeed, crews consisted of young men in the prime of sexual life, shaped by evolution to bond in serial monogamy with women of reproductive age. Of the crews who sailed into the Pacific from 1765 through 1793, 82.1 percent were between the ages of 12 and 30, and another 14.3 percent between 30 and 40. The average age of the *Bounty* crew was 26. When any of the crews arrived at the South Pacific the results, from an evolutionary point of view, were not surprising. Of the 1,556 total sailors, 437 (28 percent) got the "venereals." The *Bounty*'s infection rate was among the highest, at 39 percent (with Cook's *Resolution* and Vancouver's *Chatham* crews topping out at 57 and 59 percent, respectively).

After Bligh's men spent ten months of isolation in close quarters, he knew what would happen when they arrived at Tahiti: "The Women are handsome . . . and have sufficient delicacy to make them admired and beloved—The chiefs have taken such a liking to our People that they have rather encouraged their stay among them than otherwise, and even made promises of large possessions. Under these and many other attendant circumstances equally desirable it is therefore now not to be Wondered at . . . that a Set of Sailors led by Officers and void of connections . . . should be governed by such powerfull inducement . . . to fix themselves in the most of plenty in the finest Island in the World where they need not labour, and where the allurements of dissipation are more than equal to anything that can be conceived."

Modern neuroscience shows that the attachment bonds between men and women, especially in the early stages of a relationship, are chemical in nature and stimulate the pleasure centers of the brain in a manner resembling addictive drugs. In her book *The Oxytocin Factor*, for example, Kerstin Uvnäs Moberg shows that oxytocin is a hormone secreted into the blood by the pituitary during sex, particularly orgasm, and plays a role in pair bonding, an evolutionary adaptation for long-term care of helpless infants.

Ten months at sea weakened home attachments, that when coupled to new and powerful bonds made through sexual liaisons in Tahiti (that in some cases even led to cohabitation and pregnancy) culminated in mutiny twenty-two days after departing the island, as the men grew restless to renew those attachments (Christian, in fact, had been plotting for days to escape the *Bounty* on a raft).

Proximate causes of the mutiny may have been alcohol and anger, but the ultimate reason was evolutionarily adaptive emotions expressed non-adaptively, with irreversible consequences.

61

Unweaving the Heart

Science only adds to our appreciation for poetic beauty and experiences of emotional depth

> These good acts give pleasure, but how it happens that they give us pleasure? Because nature hath implanted in our breasts a love of others, a sense of duty to them, a moral instinct, in short, which prompts us irresistibly to feel and to succor their distresses.
>
> —Thomas Jefferson, 1814

The nineteenth-century English poet John Keats once bemoaned that Isaac Newton had "destroyed the poetry of the rainbow by reducing it to a prism." Natural philosophy, he lamented, "*will clip an Angel's wings / Conquer all mysteries by rule and line / Empty the haunted air, and gnomed mine / Unweave a rainbow.*" Keats's contemporary Samuel Taylor Coleridge concurred: "the souls of 500 Sir Isaac Newtons would go to the making up of a Shakespeare or a Milton."

Does a scientific explanation for any given phenomenon diminish its beauty or its ability to inspire poetry and experience emotions? I think not. Science and aesthetics are complementary, not conflicting; additive, not detractive. I am nearly moved to tears, for example, when I observe through my small telescope the fuzzy little patch of light that is the Andromeda galaxy. It is not just because it is lovely, but because I also understand that the photons of light landing on my retina left Andromeda 2.9 million years ago, when our ancestors were tiny-brained hominids. I am doubly stirred because it was not until 1923 that the astronomer Edwin Hubble, using the one-hundred-inch telescope on Mount Wilson, just above my home in the foothills of Los Angeles, deduced that this "nebula" was actually a distant extragalactic stellar system of immense size. He subsequently discovered that the light from most galaxies is shifted toward the red end of the electromagnetic spectrum (literally unweaving a rainbow of colors), meaning that the universe is expanding away from its explosive beginning. That is some aesthetic science. Richard Feynman expressed the complementarity of science and beauty this way:

> The beauty that is there for you is also available for me, too. But I see a deeper beauty that isn't so readily available to others. I can see the complicated interactions of the flower. The color of the flower is red. Does the fact that the plant has color mean that it evolved to attract insects? This adds a further question. Can insects see color? Do they have an aesthetic sense? And so on. I don't see how studying a flower ever detracts from its beauty. It only adds.

No less awe inspiring are recent attempts to unweave the emotions, described by Rutgers University anthropologist Helen Fisher in her book *Why We Love* (Henry Holt, 2004). Lust is enhanced by dopamine, a neurohormone produced by the hypothalamus that triggers the release of testosterone, the hormone that drives sexual desire. But love is the emotion of attachment reinforced by oxytocin, a hormone synthesized in the hypothalamus and secreted into the blood by the pituitary. In women, oxytocin stimulates birth contractions, lactation, and maternal bonding with a nursing infant. In both women and men it increases during sex and surges at orgasm, playing a role in pair bonding, an evolutionary adaptation for long-term care of helpless infants (monogamous species secrete more oxytocin during sex than polygamous species).

At the Center for Neuroeconomics Studies at Claremont Graduate University, Paul Zak posits relationships among oxytocin, trust, and economic well-being. "Oxytocin (OT) is a feel-good hormone and we find that it guides subjects' decisions even when they are unable to articulate why they are acting in a trusting or trustworthy matter," Zak explained to me. In his lab, Zak found that oxytocin increases "when a person observes that someone wants to trust him or her." Zak argues trust is among the most powerful factors affecting economic growth, and that it is vital for national prosperity that a country maximize positive social interactions among its members by ensuring a reliable infrastructure; a stable economy; and the freedom to speak, associate, and trade.

We establish trust among strangers through verification in social interactions. James K. Rilling and his colleagues at Emory University, for example, employed a functional magnetic resonance imaging (fMRI) brain scan on thirty-six subjects while they played Prisoner's Dilemma, a game in which cooperation and defection result in differing payoffs depending on what the other participants do. They found that in cooperators the brain areas that lit up were the same regions activated in response to such stimuli as desserts, money, cocaine, and beautiful faces. Specifically, the neurons most responsive were those rich in dopamine (the lust liquor that is also related to addictive behaviors) located in the anteroventral striatum in the middle of the brain—the "pleasure center." Tellingly, cooperative subjects reported increased feelings of trust toward and camaraderie with like-minded partners.

In Charles Darwin's *M Notebook*, in which he began outlining his theory of evolution shortly after returning home from his five-year voyage around the world, he penned this muse: "He who understands baboon would do more towards metaphysics than Locke." Science now reveals that love is addictive, trust is gratifying, and cooperation feels good. Evolution produced this reward system because it increased the survival of members of our social primate species. He who understands Darwin would do more toward political philosophy than Jefferson.

62

(Can't Get No) Satisfaction

The new science of happiness needs some historical perspective

Imagine you have a choice between earning $50,000 a year while other people make $25,000 or earning $100,000 a year while other people get $250,000. Prices of goods and services are the same. Which would you prefer? Surprisingly, studies show that the majority of people select the first option. As H. L. Mencken quipped, "A wealthy man is one who earns $100 a year more than his wife's sister's husband."

This seemingly illogical preference is just one of the puzzles that science is trying to solve about why happiness can be so elusive in today's world. Several recent books by researchers address the topic, but my skeptic's eye found a historian's long-view analysis to be ultimately the most enlightening.

Consider a paradox outlined by the London School of Economics economist Richard Layard in *Happiness* (Penguin, 2005), where he shows that we are no happier even though average incomes have more than doubled since 1950 and "we have more food, more clothes, more cars, bigger houses, more central heating, more foreign holidays, a shorter working week, nicer work and, above all, better health." Once average annual income is above $20,000 a head, higher pay brings no greater happiness. Why? One, our genes account for roughly half of our predisposition to be happy or unhappy, and two, our wants are relative to what other people have, not on some absolute measure. What can we do about it? Public policy can more easily remove misery than augment happiness. If we can freely exchange goods and services with each other in large and well-informed markets, and where no one affects anyone else except through the process of voluntary exchange and where contracts are enforced, the outcome will be fully efficient, that is, "everyone will be as happy as is possible without someone else being less happy."

Happiness is better equated with satisfaction than pleasure, says Emory University psychiatrist Gregory Berns in *Satisfaction* (Owl Books, 2005), because the pursuit of pleasure lands us on a never-ending hedonic treadmill that paradoxically leads to misery. "Satisfaction is an emotion that captures the uniquely human need to impart meaning to one's activities," Berns concludes. "While you might find pleasure by happenstance—winning the lottery, possessing the genes for a sunny temperament, or having the luck not to live in poverty—satisfaction can arise only by the conscious decision to do something. And this makes all the difference in the world, because it is only your own actions for which you may take responsibility and credit."

Harvard psychologist Daniel Gilbert goes deeper into our psyches in *Stumbling on Happiness* (Knopf, 2006), in which he claims, "The human being is the only animal that thinks about the future." Much of our happiness depends on projecting what *will* make us happy (instead of what actually does), and Gilbert shows that we're not very good at this forethought. Most of us imagine that variety is the spice of life, for example. But in an experiment in which subjects anticipated that they would prefer an assortment of snacks, when it actually came to eating the snacks week after week, subjects in the no-variety group said that they were more satisfied than the subjects in the variety group. "Wonderful things are especially wonderful the first time they happen," Gilbert explains, "but their wonderfulness wanes with repetition." Part of the problem lies in the fact that when we try to anticipate what will make us happy we inaccurately recall what brought us pleasure in the past. An unmarried friend of mine once confessed, "I finally had the type of weekend that married guys think single guys are having every weekend." Memory is more like an editing machine than it is a tape recorder, and as we misremember the past, we thus mispredict the future. The road to unhappiness is paved with false memories.

Gilbert also notes, with wry humor, that this habituation to even a multiplicity of wonderfulness is what economists call "declining marginal utility" and married couples call life. But if you think that an array of sexual partners adds to the spice of life, you're mistaken: according to an exhaustive study published in *The Social Organization of Sexuality* (University of Chicago Press, 1994), married people have more sex than singles . . . and more orgasms. Specifically: 40 percent of monogamous married couples have sex twice a week, compared with only 25 percent of singles, married couples are more likely to have orgasms when they do have sex, and in almost all cases men with one partner have more sex than men with multiple partners.

The historian Jennifer Michael Hecht emphasized this point in *The Happiness Myth* (Harper, 2007). Her deep and thoughtful historical perspective demonstrates just how time- and culture-dependent is all this happiness research. As she writes, "the basic modern assumptions about how to be happy are nonsense." Take sex. "If you are one of the many people who at some point in life feel sexually abnormal, note that a century ago a heterosexual married couple with cosmopolitan, secular values, having good sex three times a week, might well have felt shame and anxiety over it." By contrast, she says, "A century ago, an average man who had not had sex in three years might have felt proud of his health and forbearance, and a woman might have praised herself for the health and happiness benefits of ten years of abstinence."

Most happiness research is based on self-report data, and Hecht's point is that people a century ago would likely have answered questions on a happiness survey very differently than they do today. Applying the methods of science to finding happiness is music to the ears of someone like me, who has been accused of excessive scientism, but after reading a number of these books I was most moved toward enlightenment by the analysis of a historian whose long-view perspective leads me to conclude that much of this science is time-dependent.

According to the best science today, then, what brings happiness? Social bonds (coworkers, friends, marriage), trust in people (friends, family, strangers), trust in society (the economy, justice, government), religion and spirituality (prayer, meditation, positive psychology), and prosocial behavior (helping the poor, volunteering). What brings unhappiness? Divorce, unemployment, loss of status, depression, poverty. (As Woody Allen jibed: "Money is better than poverty, if only for financial reasons.")

To understand happiness, we need both history and science.

IX

EVOLUTION AND CREATIONISM

63

The Gradual Illumination of the Mind

The advance of science, not the demotion of religion, will best counter the influence of creationism

In one of the most starkly honest and existentially penetrating statements ever made by a scientist, Richard Dawkins concluded that "the universe we observe has precisely the properties we should expect if there is, at bottom, no design, no purpose, no evil and no good, nothing but blind, pitiless indifference."

Facing such a reality perhaps we should not be surprised at the results of a 2001 Gallup poll confirming that 45 percent of Americans agree with the statement "God created human beings pretty much in their present form at one time within the last 10,000 years or so," 37 percent prefer a blended belief that "human beings have developed over millions of years from less advanced forms of life, but God guided this process," and a paltry 12 percent accept the standard scientific theory that "human beings have developed over millions of years from less advanced forms of life, but God had no part in this process."

In a forced binary choice between the "theory of creationism" and the "theory of evolution," 57 percent chose creationism against only 33 percent for evolution (10 percent said they were "unsure"). One explanation for these findings can be seen in additional results showing that only 33 percent of Americans think that the theory of evolution is "well supported by evidence," while slightly more (39 percent) believe that it is not well supported, and that it is "just one of many theories." A quarter surveyed said they didn't know enough to say and, shedding some light on the problem, only 34 percent considered themselves to be "very informed" about evolution.

Although such findings are disturbing, truth in science is not determined democratically. It does not matter whether 99 percent or only 1 percent of the public believes a theory. It must stand or fall on the evidence, and there are few theories in science that are more robust than the theory of evolution. The preponderance of evidence from numerous converging lines of inquiry (geology, paleontology, zoology, botany, comparative anatomy, genetics, biogeography, etc.) all independently point to the same conclusion: evolution happened. The nineteenth-century philosopher of science William Whewell called this process of independent lines of inquiry converging together to a conclusion a "consilience of inductions." I call it a "convergence of evidence." Whatever you call it, it is how historical events are proven.

The reason we are experiencing this peculiarly American phenomenon of *evolution denial* (the doppelgänger of *Holocaust denial*, using the same techniques of rhetoric and debate—see my book *Why People Believe Weird Things* for a comparison) is that a small but vocal minority of religious fundamentalists misread the theory of evolution as a challenge to their deeply held religious convictions. Given this misunderstanding, their response is to attack the theory. It is no coincidence that almost without exception all of the evolution deniers are Christians who believe that if God did not personally intervene in the development of life on Earth then they have no basis for belief, morality, and the meaning of life. Clearly for some, much is at stake in the findings of science.

Since the Constitution prohibits public schools from promoting any particular brand of religion, this has led to the oxymoronic movement known as "creation-science," or, in its more recent incarnation, "Intelligent Design" (ID), where ID (aka God) miraculously intervenes just in the places where science has yet to offer a comprehensive explanation for a particular phenomenon. (ID used to control the weather, but now that we understand it He has moved on to more difficult problems, such as the origins of DNA and cellular life. Once these problems are mastered then

ID will no doubt find even more intractable conundrums.) Thus IDers would have us teach children nonthreatening theories of science, but when it comes to the origins of life and certain aspects of evolution, children are to learn that "ID did it." I fail to see how this is science, or what it is, exactly, that IDers hope will be taught in these public schools. "ID did it" makes for a rather short semester.

By contrast, a scientist would want to know *how* ID did it. In eschewing all attempts to provide a naturalistic explanation for the phenomena under question, IDers have abandoned science altogether. Yet they want the respectability that science brings in our culture, so they do theology and call it science.

To counter the nefarious influence of the ID creationists we need to employ a proactive strategy of science education and evolution explanation. It is not enough to argue that creationism is wrong; we must also show that evolution is right. The theory's founder, Charles Darwin, knew this when he reflected, "It appears to me (whether rightly or wrongly) that direct arguments against Christianity and theism produce hardly any effect on the public; and freedom of thought is best promoted by the gradual illumination of men's minds which follows from the advance of science."

64

Vox Populi

The voice of the people reveals why evolution remains controversial

Sigmund Freud famously (and immodestly) once observed that humanity has suffered three great intellectual shocks: the first when Copernicus placed us on "a tiny speck in a world-system of a magnitude hardly conceivable," the second when Darwin "robbed man of his peculiar privilege of having been specially created, and relegated him to a descent from the animal world," and the third when Freud's own theory proved "to the 'ego' of each one of us that he is not even master of his own house, but that he must remain content with the veriest scraps of information about what is going on unconsciously in his own mind."

Although Copernicus and Freud have left the public stage of debate and disputation (Copernicus because of complete acceptance and Freud because of near universal rejection), Darwin remains (at least in America) mired in a slurry of religious and political controversy. This fact was brought to light for me in the overwhelming response to my February 2002 column "The Gradual Illumination of the Mind," on evolution and Intelligent Design creationism. Where I typically receive about a dozen letters a month, for this one no less than 134 were submitted (117 men, 4 women, 13 unknowns—a ratio equivalent to the magazine's gender split). Most were niggling over my position that creationism is religion masquerading as science, and that evolution is one of the best supported theories in the history of science.

When I first started my column in *Scientific American* I found reading critical letters mildly disconcerting until I hit upon the idea that these are a form of data to be mined for additional information on what people believe and why. Conducting a content analysis of all 134 letters, I discovered a pattern within the cacophonous chaos. Initially I read through them all quickly, coding them into about two dozen one-line categories that summed up the reader's point. I then coalesced these into six taxonomic classes, and reread all the letters carefully, placing each into one or more of the six (many readers made more than one point, giving a total of 163 ratings from which the percentages below were derived). Results:

1. Evolution is true/creationism is false/science > religion	7%
2. Evolution is God's method of creation/science=religion	12%
3. Evolution is false/creationism is true/religion > science	16%
4. Evolution requires faith to believe/science is a religion	17%
5. Intelligent Design is true/life is too complex to have evolved	23%
6. Noncommittal/science ≠religion/Alternate theories of evolution/religion	25%

Excerpts from the letters illuminate each taxon. (Although most were friendly and reasonable, one fellow opined that my column "could have been written in 1939 by a Nazi," while another said that "Michael Shermer must not only be a sceptic but also a stupid in the 3rd degree the way he talks about 'Intelligent Design.'") Disturbingly, only 7 percent agreed with me about the veracity of evolution (and the emptiness of creationism), with one reader going so far as to claim that "the defenders of science behave too well. No amount of evolution education will counter the deliberate, sly, selective ignorance of creation 'science.'"

Nearly double that number argued that evolution is God's method of creating life, such as one correspondent who agreed "that evolution is right—but still I see GOD in the will and cunning intention in the genetic system of all living organisms and in the system and order present in the laws of nature. Seeing all the diversity in the methods of camoflage in animals and plants for an example, I know that there is a will behind it."

Another reader sees creation and evolution "as complementary to each other. Put simply, since all parts of the universe follow intelligible law as educed through human intelligence, and such a law is a principle or cause, it follows that the universe as a whole must be the effect of the operation of a singular all-encompassing Principle."

Critics of evolution in the third taxon hauled out an old canard every evolutionary biologist has heard: "I want to point out that evolution is only a theory." And: "To my knowledge evolution is just a theory that has never been put to the test successfully and is far from being conclusive."

That evolution requires faith to believe (the fourth category) found many adherents among readers, such as these: "There are so many vast chasms that evolutionists paint over with broad strokes, they act as if their faith is fact as often as a creationist." Or: "On my view, a key part of being a rational skeptic is consistent dedication to the standards and methods of critical thinking and logic. In his zeal to defend his faith in evolutionary theory, Dr. Shermer violates those standards." My favorite letter in this class echoed a standard refrain we hear often at *Skeptic* magazine about inadequate or misplaced skepticism (with a cc list that included "Pres. George W. Bush, V.P. Dick Cheney, and The Members of The US Congress, American Academy of Science; Dr. Dean Edell, America's Doctor; Dr. Laura, America's Jewish Mother"!): "I applaud your SKEPTICISM when it comes to Creationism and Astrology and 'Psychic Phenomena'; but how can you be so THICK HEADED when it comes to the GLARING WEAKNESSES of Darwinian Evolution??? Honestly, you come across as both a 'brainwashed apologist' and a 'High School Cheerleader' for Darwinian Evolution."

Charles Darwin, he's our man. If he can't do it no one can.

The penultimate taxon was that Intelligent Design creationism must be true because life is simply too complex to be explained by evolution (corresponding to the results of my study on religious beliefs, reported in my book *How We Believe*, that the number one reason people say they believe in God is because of the apparent good design of the universe and life). For example: "ID theorists also see a variety of factors, constants, and relationships in the construction of the universe which are so keenly well-adjusted to the existence of matter and life that they find it impossible to deny the implication of intelligent purpose in those factors.

Materialists see the same thing and wave their hands vaguely and mutter mystical phrases about 'Anthropic Principles.' What the materialist calls the anthropic principle, the IDer calls the Designer."

Intriguingly, the greatest number of responses fell into a noncommittal position where readers expounded on the relationship of science and religion, often presenting their own theories of evolution and creation as alternatives to the models under discussion. For example: "Evolution is not a theory. It is an analytic approach. There are three elements of science: operation, observation, and model. An observation is the result of applying an operation, and a model is chosen for its utility in explaining, predicting, and controlling observations, balanced against the cost of using it." And: "There is nothing that scientists have ever discovered, or could ever discover, that can prove or disprove the existence of God. The Bible is a tool for the illumination of the heart, not the revelation of observable facts. Thus there is no conflict between the Bible and science—there is even an amazing synergy between the two—when each is kept in its proper place."

It has been my experience that correspondents in this final category, like questioners in the Q & A sessions of lectures I present at colleges and universities, are less interested in my opinion and more intent on launching their own ideas into the cultural ether. With no subject is this more apparent than for evolution; it is here we face the ultimate questions of genesis and exodus: where did we come from and where are we going? No matter how you answer those questions, facing them with courage and intellectual honesty will bring you closer to the creation itself.

65

The Fossil Fallacy

Creationists' demand for "just one transitional fossil" reveals a deep misunderstanding of science

The nineteenth-century English social scientist Herbert Spencer made this prescient observation: "Those who cavalierly reject the Theory of Evolution, as not adequately supported by facts, seem quite to forget that their own theory is supported by no facts at all." A century later nothing has changed. When I debate creationists they present not one fact in favor of creation and instead demand "just one transitional fossil" that proves evolution. When I do (for example, *Ambulocetus natans*, a transitional fossil between the land mammals *Mesonychids* and the marine mammals *Archaeocetes*, an ancestor of modern whales) they respond that there are now *two gaps* in the fossil record.

This is a clever debate retort, but it reveals a deep error in epistemology that I call the *Fossil Fallacy*: the belief that a single "fossil"—one bit of data—constitutes proof of a multifarious process or historical sequence. In fact, proof is derived through a convergence of evidence from numerous lines of inquiry—multiple inductions from multiple data sets—all of which point to an unmistakable conclusion.

(As a rhetorical tactic intended to confound Holocaust historians accustomed to playing by the accepted rules of scholarship, Holocaust deniers demand "just one proof" of the veracity of the central tenets of the Shoah. For example, they ask: Where are the Zyklon-B gas pellet induction holes in the roof of the gas chamber in Krema II at Auschwitz-Birkenau? No holes, no Holocaust, they claim, even codifying this conclusion on a T-shirt. The fallacy here is assuming that the Holocaust is a single event that can be "proven" by a single piece of data. This is the same error that evolution deniers make in demanding "just one transitional fossil" that proves evolution happened. Just as the Holocaust was thousands of events that occurred in thousands of places and is proved [reconstructed] through thousands of historical facts, evolution is a process and historical sequence proved through thousands of bits of data from numerous fields of science that together give us a rich portrait of the history of life.)

We know evolution happened not because of transitional fossils such as *Ambulocetus natans*, but because of the convergence of evidence from such diverse fields as geology, paleontology, biogeography, comparative anatomy and physiology, molecular biology, genetics, and many more. No single discovery from any of these fields denotes proof of evolution, but together they converge to reveal that life evolved in a specific sequence by a particular process.

The problem with fossils is threefold: (1) most dead organisms do not fossilize; (2) those that do fossilize are unlikely ever to be found; and (3) transitional fossils are rare because the most common speciation process (allopatric speciation) is relatively rapid and happens in small founder populations that break off from larger parent populations, with fewer numbers of individuals available to fossilize.

One of the finest compilations of evolutionary data and theory since Charles Darwin's *The Origin of Species* is Richard Dawkins's magnum opus *The Ancestor's Tale: A Pilgrimage to the Dawn of Evolution* (Houghton Mifflin, 2004), 673 pages of convergent science recounted with literary elegance. Dawkins traces innumerable "transitional fossils" (what he calls "concestors"—the "point of rendezvous" of the last common ancestor shared by a set of species) from *Homo sapiens* back four billion years to the origin of heredity and the emergence of evolution. No one concestor proves that evolution happened, but together they reveal a majestic story of process over time.

Consider the tale of the dog. With so many breeds of dogs popular for so many thousands of years, one would think that there would be an abundance of transitional fossils providing paleontologists with copious data from which to reconstruct their evolutionary ancestry. In fact, according to Jennifer A. Leonard of the National Museum of Natural History in Washington, DC, "the fossil record from wolves to dogs is pretty sparse." Then how do we know whence dogs evolved? In the November 22, 2002, issue of *Science*, Leonard and her colleagues report that mitochondrial DNA (mtDNA) data from early dog remains "strongly support the hypothesis that ancient American and Eurasian domestic dogs share a common origin from Old World gray wolves."

In the same issue, Peter Savolainen from the Royal Institute of Technology in Stockholm and his colleagues note that the fossil record is problematic "because of the difficulty in discriminating between small wolves and domestic dogs," but their study of mtDNA sequence variation among 654 domestic dogs from around the world "points to an origin of the domestic dog in East Asia ~15,000 yr B.P." from a single gene pool of wolves.

Finally, in this same issue, Brian Hare from Harvard and his colleagues describe the results of their study in which they found that domestic dogs are more skillful than wolves at using human communicative signals indicating the location of hidden food, but that "dogs and wolves do not perform differently in a non-social memory task, ruling out the possibility that dogs outperform wolves in all human-guided tasks." Therefore, "dogs' social-communicative skills with humans were acquired during the process of domestication."

Although no single fossil proves that dogs came from wolves, archaeological, morphological, genetic, and behavioral "fossils" converge to reveal the concestor of all dogs to be the East Asian wolf. The tale of human evolution is revealed in a similar manner (although here we do have an abundance of fossil riches), as it is for all concestors in the history of life. We know evolution happened because innumerable bits of data from myriad fields of science conjoin to paint a rich portrait of life's pilgrimage.

Update: A 2013 paper published in PLOS (Public Library of Science) *by Anna Druzhkova and Olaf Thalmann of the Russian Academy of Sciences presented the results of DNA analysis from the bones of a*

33,000-year-old Pleistocene dog from the Altai region of central Asia (the second oldest dog specimen ever found), showing that this dog "is more closely related to modern dogs and prehistoric New World canids than it is to contemporary wolves." The authors caution that their results are based on a single specimen, but if it holds up it doubles the age of the origin of the modern dog. http://bit.ly/1B8U759

66

Rumsfeld's Wisdom

Where the known meets the unknown is where science begins

At a February 12, 2002, news briefing, Secretary of Defense Donald Rumsfeld explained the limitations of intelligence reports: "There are known knowns. There are things we know we know. We also know there are known unknowns. That is to say, we know there are some things we do not know. But there are also unknown unknowns, the ones we don't know we don't know."

Rumsfeld's logic may be tongue-twisting, but his epistemology was sound enough that he was quoted twice at the World Summit on Evolution in June 2005, hosted by the Universidad San Francisco de Quito and held on the Galápagos island of San Cristóbal, where Darwin began his explorations. Rumsfeld's wisdom was first invoked by University of California at Los Angeles paleobiologist William Schopf, who, in a commentary on a lecture on the origins of life, asked: "What do we know? What are the unsolved problems? What have we failed to consider?"

Creationists and outsiders often mistake the latter two categories for signs that the theory of evolution is in trouble, or that contentious debate between what we know and do not know means that the theory is false, or that science is a cozy club in which meetings are held to reinforce the party line. Wrong. The summit revealed a scientific theory rich in data and theory, as well as controversy and disputation over the known and unknown.

For example, Schopf began with the known: "We know the overall sequence of life's origin, from CHONSP [carbon, hydrogen, oxygen, nitrogen, sulfur, phosphorus], to monomers, to polymers, to cells; we know that the origin of life was early, microbial, and unicellular; and we know that an RNA world preceded today's DNA-protein world. We do not know the precise environments of the early Earth in which these events occurred; we do not know the exact chemistry of some of the important chemical reactions that led to life; and we do not have any knowledge of life in a pre-RNA world." As for what we have failed to consider, Schopf noted a problem with what he called "the pull of the present"—it is extremely difficult to model the early Earth's atmosphere and the biochemistry of early life because we are so accustomed to conditions today, adding that Darwin's "warm little pond" may or may not explain life's origin.

Rumsfeld's heuristic was summoned again at the end of the conference by University of Georgia evolutionary biologist Patricia Gowaty, in response to Stanford University biologist Joan Roughgarden, who declared that Darwin's theory of sexual selection is wrong in its claim that females choose mates who are more attractive and well armed. "People are surprised to learn how much sex animals have for purely social reasons, and how many species have sex-role reversal in which the males are drab and the females are colorfully ornamented and compete for the attention of males." Gowaty noted that Roughgarden is right in identifying the exceptions to Darwin's theory and that there are many unknowns, but she added that since Darwin's time much has been learned about mate selection and competition that is well known.

Between these Rumsfeldian bookends, scientific skepticism was rampant. University of Massachusetts biologist Lynn Margulis said "neo-Darwinism is dead" because "random changes in DNA alone do not lead to speciation. Symbiogenesis—the appearance of new behaviors, tissues, organs, organ systems, physiologies, or species as a result of symbiont interaction—is the major source of evolutionary novelty in eukaryotes: animals, plants, and fungi." University of California at Berkeley paleoanthropologist Timothy White suggested that his colleagues have engaged in far too much species splitting in classifying fossil hominids. American Museum of Natural History paleontologist Niles Eldredge explained how punctuated equilibrium—the idea that long periods of species stability are punctuated by rapid bursts of speciation—better accounts for the fossil record than gradualism, which posits slow and steady evolutionary change.

As I was in attendance to inform about Intelligent Design theory, I had a nightmarish thought: creationists could have a field day yanking quotes out of context while listening to a room full of evolutionary biologists arguing over specific issues. In point of fact, such debates are all *within* evolutionary theory, not between evolutionary theory and something else. This boundary between the known and the unknown is where science flourishes.

67

It's Dogged as Does It

Retracing Darwin's footsteps in the Galápagos shatters a myth but reveals how revolutions in science actually evolve

Among the many traits that made Charles Darwin one of the greatest minds in science was his pertinacious personality. Facing a daunting problem in natural history Darwin would obstinately chip away at it until its secrets relented. His apt description for this disposition came from an 1867 Anthony Trollope novel in which one of the characters opined: "There ain't naught a man can't bear if he'll only be dogged. . . . It's dogged as does it." Darwin's son Francis recalled his father's temperament: "Doggedness expresses his frame of mind almost better than perseverance. Perseverance seems hardly to express his almost fierce desire to force the truth to reveal itself."

Historian of science Frank J. Sulloway, of the University of California–Berkeley, has highlighted Darwin's "dogged" genius in his own tenacious efforts to force the truth of how the theory of evolution was actually pieced together by Darwin. The iconic myth is that Darwin became an evolutionist in the Galápagos when he discovered natural selection operating on finch beaks and tortoise carapaces, each species uniquely adapted by food type or island ecology. The myth is ubiquitous, appearing in everything from biology textbooks to travel brochures, the latter of which inveigle potential customers to visit the mecca of evolutionary theory and walk in the footsteps of St. Darwin the Divine.

In June 2004, Sulloway and I did just that, spending a month retracing some of Darwin's fabled footfalls. Sulloway is one sagacious scientist, but I had no idea he was such a dogged field explorer until we hit the lava on San Cristóbal to reconstruct Darwin's explorations there. Doggedness is the watchword here, for with a sweltering equatorial sun and almost no fresh water, it isn't long before seventy-pound water-loaded packs begin to buckle knees and strain backs. Add hours of daily bushwhacking through dry, dense, scratchy vegetation and the romance of fieldwork quickly fades.

Yet the harder it got, the more tenacious Frank became. He actually seemed to enjoy the misery, and this gave me a glimpse into Darwin's doggedness. At the end of one particularly grueling climb through a moonscape-like area Darwin called the "craterized district" of San Cristóbal, we collapsed in utter exhaustion, muscles quivering and sweat pouring off our hands and faces; Darwin termed a similar excursion "a long walk."

Death permeates these islands. Animal carcasses are scattered hither and yon. The vegetation is coarse and scrappy. Dried and shriveled cacti trunks dot the bleak lava landscape that is so broken with razor-sharp edges that moving across it is glacially slow. Many people have died, from stranded sailors of centuries past to wanderlust tourists of recent years. Within days I had a deep sense of isolation and fragility. Without the protective blanket of civilization none of us is far from death. With precious little water and even less eatable foliage, organisms eke out a precarious living, their adaptations to this harsh environment selected over millions of years. These critters are hanging on by the skin of their adaptive radiations. A lifelong observer of and participant in the creation-evolution controversy, I was struck by how clear it is in these islands: creation by Intelligent Design is absurd. Why, then, did Darwin depart the Galápagos a creationist?

The Darwin Galápagos legend is emblematic of a broader myth that science proceeds by select eureka discoveries followed by sudden revolutionary revelations, where old theories fall before new facts. Not quite. Paradigms power percepts. Nine months after departing the Galápagos, Sulloway discovered, Darwin made the following entry in his ornithological catalog about his mockingbird collection: "When I see these Islands in sight of each other, & possessed of but a scanty stock of animals, tenanted by these birds, but slightly differing in structure & filling the same place in Nature, I must suspect they are only varieties." Similar *varieties* of fixed kinds, not *evolution* of separate species. Darwin was still a creationist! This explains why Darwin didn't even bother to record the island locations of the few finches he collected (and in some cases mislabeled), and why, as Sulloway has pointed out, these now-famous birds were never specifically mentioned in *Origin of Species*.

Darwin similarly botched his tortoise observations, as he recalled later from a conversation he had while in the islands with the vice governor Nicholas O. Lawson, who explained that for the tortoises "he could with certainty tell from which islands any one was brought. I did not for some time pay sufficient attention to this statement, and I had already partially mingled together the collections from two of the islands." Worse, as Sulloway recounts humorously, Darwin and his mates ate the remaining tortoise data on the voyage home. As Darwin confessed, "I never dreamed that islands, about fifty or sixty miles apart, and most of them in sight of each other, formed of precisely the same rocks, placed under a quite similar climate, rising to a nearly equal height, would have been differently tenanted."

Darwin departed the Galápagos in October 1835 but did not become an evolutionist until March 1837, and continued to refine his theory until *Origin of Species* was published in 1859. And, following in Darwin's conceptual as well as literal footsteps, Sulloway derived this conclusion not from an abrupt discovery in the Galápagos, but from a careful analysis of Darwin's data back in England. Through careful analysis of Darwin's notes and journals, Sulloway dates Darwin's acceptance of the theory of evolution to the second week of March 1837, after a meeting Darwin had with the eminent ornithologist John Gould, who had been studying his Galápagos bird specimens. With access to museum ornithological collections from areas of South America that Darwin had not visited, Gould corrected a number of taxonomic errors Darwin had made (such as labeling two finch species a "Wren" and "Icterus"), and pointed out to him that although the land birds in the Galápagos were endemic to the islands, they were notably South American in character.

Darwin left the meeting with Gould, Sulloway concludes, convinced "beyond a doubt that transmutation must be responsible for the presence of similar but distinct species on the different islands of the Galápagos group. The supposedly immutable 'species barrier' had finally been broken, at least in Darwin's own mind." That July 1837 Darwin opened his first notebook on the subject, *Transmutation of Species*, in which he noted, "Had been greatly struck from about Month of previous March on character of S. American fossils—and species on Galapagos Archipelago. These facts origin (especially latter) of all my views." By 1845 Darwin was confident enough in his data to theorize on the deeper implications of the Galápagos in the second edition of his *Journal of Researches*, which includes one of the great one-liners in science history: "The archipelago is a little world within itself, or rather a satellite attached to America, whence it has derived a few stray colonists, and has received the general character of its indigenous productions. . . . Hence both in space and time, we seem to be brought somewhat near to that great fact—that mystery of mysteries—the first appearance of new beings on this earth."

For a century and a half Darwin's theory has doggedly explained more disparate facts of nature than any other in the history of biology; the process itself is equally dogged, as Darwin explained: "It may be said that natural selection is daily and hourly scrutinising, throughout the world, every variation, even the slightest; rejecting that which is bad, preserving and adding up all that is good; silently and insensibly working, whenever and wherever opportunity offers."

Doggedly so.

68

Darwin on the Right

Why Christians and conservatives should accept evolution

According to a 2005 Pew Research Center poll, 70 percent of evangelical Christians believe that living beings have always existed in their present form, compared with 32 percent of Protestants and 31 percent of Catholics; politically, 60 percent of Republicans are creationists, whereas only 11 percent accept evolution, compared with 29 percent of Democrats who are creationists and 44 percent who accept evolution. A 2005 Harris poll found that 63 percent of liberals but only 37 percent of conservatives believe that humans and apes have a common ancestry, and that those with a college education, ages eighteen to fifty-four, from the Northeast and West are more likely to accept evolution, whereas those without a college degree, age fifty-five and older, and from the South are more likely to believe in creationism.

What these figures confirm for us is that there are religious and political reasons for rejecting evolution. Can one be a conservative Christian and a Darwinian? Yes. Here's how and why.

1. *Evolution fits well with good theology.* Christians believe in an omniscient and omnipotent God. What difference does it make *when* God created the universe—10,000 years ago or 10,000,000,000 years ago?

The glory of the creation commands reverence regardless of how many zeros are in the date. And what difference does it make *how* God created life—spoken word or natural forces? The grandeur of life's complexity elicits awe regardless of what creative processes were employed. Christians (indeed, all faiths) should embrace modern science for what it has done to reveal the magnificence of the divinity in a depth and detail unmatched by ancient texts.

2. *Creationism is bad theology.* The watchmaker God of Intelligent Design creationism is delimited to being a garage tinkerer piecing together life out of available parts. This God is just a genetic engineer slightly more advanced than we are. An omniscient and omnipotent God must be above such human-like constraints. As the Protestant theologian Langdon Gilkey wrote, "the Christian idea, far from merely representing a primitive anthropomorphic projection of human art upon the cosmos, systematically repudiates all direct analogy from human art." Calling God a watchmaker is belittling.

3. *Evolution explains original sin and the Christian model of human nature.* As a social primate we evolved within-group amity and between-group enmity. By nature, then, we are cooperative and competitive, altruistic and selfish, greedy and generous, peaceful and bellicose; in short, good and evil. Moral codes and a society based on the rule of law are necessary to accentuate the positive and attenuate the negative sides of our evolved nature.

4. *Evolution explains family values.* The following characteristics are the foundation of families and societies and are shared by humans and other social mammals: attachment and bonding, cooperation and reciprocity, sympathy and empathy, conflict resolution and peacemaking, community concern and reputation anxiety, and response to group social norms. As a social primate species we evolved morality to enhance the survival of both family and community; subsequently, religions designed moral codes based on our evolved moral natures.

5. *Evolution explains specific Christian moral precepts.* Much of Christian morality has to do with human relationships, most notably truth telling

and marital fidelity, because violation of these causes a severe breakdown in trust, which is the foundation of family and community. Evolution explains why. We evolved as pair-bonded primates, and adultery is a violation of a monogamous relationship. Likewise, truth telling is vital for trust in social relations, so lying is a sin.

6. *Evolution explains conservative free market economics.* Charles Darwin's *natural selection* is precisely parallel to Adam Smith's *invisible hand.* Darwin showed how complex design and ecological balance were unintended consequences of individual competition among organisms. Smith showed how national wealth and social harmony were unintended consequences of individual competition among people. Nature's economy mirrors society's economy. Both are designed from the bottom up, not the top down.

By providing a scientific foundation for the core values shared by most Christians and conservatives, the theory of evolution should be embraced. The senseless conflict between science and religion must end now, or else, as the Book of Proverbs (11:29) warned, "*He that troubleth his own house shall inherit the wind.*"

X

SCIENCE, RELIGION, MIRACLES, AND GOD

69

Digits and Fidgets

Is the universe fine-tuned for life?

There was a young fellow from Trinity
Who took [the square root of infinity]
But the number of digits
Gave him the fidgets;
he dropped Math and took up Divinity.

In this limerick physicist George Gamow dealt with the paradox of a finite being contemplating infinity by passing the buck to theologians.

In an attempt to prove that the universe was intelligently designed, religion has lately been fidgeting with the "fine tuning" digits of the cosmos. "It is not only man that is adapted to the universe," physicists John Barrow and Frank Tipler proclaim in their 1986 book *The Anthropic Cosmological Principle.* "The universe is adapted to man. Imagine a universe in which one or another of the fundamental dimensionless constants of physics is altered by a few percent one way or the other? Man could never come into being in such a universe."

The Templeton Foundation even grants cash prizes for such "progress in religion." In 2002 mathematical physicist and Anglican priest John C. Polkinghorne was given $1 million for his "treatment of theology as a natural science" and who "has invigorated the search for interface between science and religion." In 1997 physicist Freeman Dyson took home $964,000 for such works as his 1979 book *Disturbing the Universe*, in which he writes, "As we look out into the universe and identify the many accidents of physics and astronomy that have worked to our benefit, it almost seems as if the universe must in some sense have known that we were coming."

Mathematical physicist Paul Davies won the 2000 Templeton Prize for such observations as those made in his 1999 book *The Fifth Miracle*: "If life follows from [primordial] soup with causal dependability, the laws of nature encode a hidden subtext, a cosmic imperative, which tells them: 'Make life!' And, through life, its by-products: mind, knowledge, understanding. It means that the laws of the universe have engineered their own comprehension. This is a breathtaking vision of nature, magnificent and uplifting in its majestic sweep. I hope it is correct. It would be wonderful if it were correct." Indeed, it would be wonderful. But not any more wonderful than if it was not correct.

Even atheist Stephen Hawking sounds like a supporter of Intelligent Design when he writes, "Why is the universe so close to the dividing line between collapsing again and expanding indefinitely? If the rate of expansion one second after the big bang had been less by one part in 10^{10}, the universe would have collapsed after a few million years. If it had been greater by one part in 10^{10}, the universe would have been essentially empty after a few million years. In neither case would it have lasted long enough for life to develop. Thus one either has to appeal to the anthropic principle or find some physical explanation of why the universe is the way it is."

One explanation is that our universe is not the only one. We may live in a multiverse in which our universe is just one of many bubble universes all with different laws of nature. Those bubble universes whose parameters are most likely to give rise to life occasionally generate complex life with brains big enough to achieve consciousness and to conceive of such concepts as God and cosmology, and to ask such questions as Why?

Another explanation can be found in the properties of *self-organization* and *emergence*. Water is an emergent property of a particular arrangement of hydrogen and oxygen molecules, just as consciousness is a self-organized emergent property of billions of neurons. The evolution of complex life is an emergent property of simple life: prokaryote cells self-organized into eukaryote cells, which self-organized into multicellular organisms, which self-organized into . . . and here we are.

Self-organization and emergence arise out of *complex adaptive systems* that grow and learn as they change. As a complex adaptive system the cosmos is one giant autocatalytic (self-driving) feedback loop that generates such emergent properties as life. We can even think of self-organization as an emergent property, and emergence as a form of self-organization. How recursive. Complexity is so simple it can be put on a bumper sticker: LIFE HAPPENS.

If life on Earth is unique, or at least exceptionally rare (and in either case certainly not inevitable), how special is our fleeting mayfly-like existence; how important it is that we make the most of our lives and our loves; how critical it is that we work to preserve not only our own species, but also all species and the biosphere itself. Whether the universe is teeming with life or we are alone, whether our existence is strongly necessitated by the laws of nature or is highly contingent and accidental, whether there is more to come or this is all there is, we are faced with a worldview that is breathtaking and majestic in its sweep across time and space.

70

Remember the 6 Billion

For millennia we have raged against the dying of the light.
Can science save us from that good night?

Between now and the year 2123 a tragedy of Brobdingnagian proportions will befall humanity, causing the death of more than 6 billion people. I'm serious.

According to Carl Haub, a demographer at the Population Reference Bureau of the US Census in Washington, DC, between 50,000 BC and 2002, 106,456,367,669 people were born. The Earth's population is now 7,290,289,811 (in 2015). Of the 100 billion people who came before us, every one of them has died. To the extent that the past is the key to the present—and the future—that means that within the next 120 years (the maximum life potential) more than 6 billion humans will suffer the same fate. And there is not a damn thing we can do about it. Or is there?

Until the late twentieth century, when science took up the cause, the only recourses anyone had in the face of this reality were prayer and poetry. The seventeenth-century English poet John Donne, for example, knew all too well for whom the bell tolls (his wife died at age thirty-three, after giving birth to the twelfth of their children, five of whom died), decrying, *"Death be not proud, though some have called thee/Mighty and dreadful, for, thou art not so."*

Today we are being offered scientistic alternatives, if not for immortality itself, then for longevity of biblical proportions. All have some basis in science, but none has achieved anything like scientific confirmation and thus fall into the realm of either borderlands science or pseudoscience.

Virtual immortality. According to Tulane University physicist Frank Tipler, in the far future we will all be resurrected in a virtual reality whose memory capacity is 10 to the power of 10^{123} (a 1 followed by 10^{123} zeros). If the virtual reality were real enough, it would be indistinguishable from our reality. Boot me up, Scotty. . . .

Genetic immortality. Oh, those pesky telomeres at the end of chromosomes that prevent cells from replicating indefinitely. If only we could genetically reprogram them to be like cancer cells. Alas, this is no solution because biological systems are so complex that fixing any one component does not address all the others that play a role in aging.

Cryonics immortality. Freeze—wait—reanimate. It sounds good in theory, but you're still a corpsicle. Don't forget to pay the electric bill in the meantime.

Replacement immortality. First we replace our organs, then our cells, then our molecules, nano-a-nano, eventually exchanging protein (flesh) for something more durable, such as silicon. You can't tell the difference, can you?

Lifestyle longevity. Here is something we can implement today, which means the hucksters are out in force offering every elixir under and including the sun to extend human life. To cut to the chase, S. Jay Olshansky, Leonard Hayflick, and Bruce A. Carnes, three of the world's leading scientists on aging, stated unequivocally in *Scientific American* ("No Truth to the Fountain of Youth," June 2002), "no currently marketed intervention—none—has yet been proved to slow, stop or reverse human aging, and some can be downright dangerous."

It has never been proven, for example, that antioxidants—taken as supplements to counter the deleterious effects of free radicals on cells—attenuate aging. In fact, free radicals are necessary for cellular physiology. Hormone replacement therapy, another popular antiaging nostrum, is effective for some short-term shortcomings such as loss of muscle mass and strength in older men and postmenopausal women. But the long-term negative side effects are still unknown and the slowing of the aging process unproven.

As a lifelong cyclist I am pleased to report that diet and exercise are tried and true methods to increase your life span. These, along with modern medical technologies and sanitation practices, have nearly doubled the average life expectancy over the past century. Unfortunately, this just means that more of us will get closer to the outer wall of 120 years before finally and inexorably succumbing to the way of all flesh, no matter how you may personally express Dylan Thomas's sentiment now annealed into Western literature:

Do not go gentle into that good night
Rage, rage against the dying of the light.

Rage all you like, but remember the 6 billion—and the 100 billion before. Until science finds a solution to prolonging the duration of healthy life, we should instead rave about the time we have, however fleeting.

71

God's Number Is Up

Among a heap of books claiming that science proves God's existence emerges one that computes a probability of 67 percent

In his 1916 poem "A Coat," William Butler Yeats rhymed:

I made my song a coat
Covered with embroideries
Out of old mythologies
From heel to throat.

Read religion for song, and science for coat, and we have a close approximation to the deepest flaw in the science and religion movement, as revealed in Yates's denouement:

But the fools caught it,
Wore it in the world's eyes
As though they'd wrought it.
Song, let them take it
For there's more enterprise
In walking naked.

Naked faith is what religious enterprise was always about, until science became the preeminent system of natural verisimilitude, tempting the faithful to employ her wares in the practice of preternatural belief. Although most efforts in this genre offer little more than scientistic cant and religious blather, a few require a response from the magisterium of science if for no other reason than to protect the magisterium of religion—if faith is tethered to science, what happens when the science changes? One of the most innovative works I have encountered in this genre is *The Probability of God*, by physicist Stephen D. Unwin, whose early research on quantum gravity showed him that the universe is probabilistic, and whose later work in risk analysis led him to this ultimate computation.

Unwin rejects most scientific attempts to prove the divine—such as the anthropic principle and Intelligent Design—concluding that this "is not the sort of evidence that points in either direction, for or against." Instead, Unwin employs "Bayesian probabilities," a statistical method devised by the eighteenth-century Presbyterian minister and mathematician the Reverend Thomas Bayes. Unwin begins with a 50 percent probability that God exists (because 50/50 represents "maximum ignorance"), then applies a modified Bayesian theorem:

$$P_{after} = \frac{P_{before} \times D}{P_{before} \times D + 100\% - P_{before}}$$

The probability of God's existence after the evidence is considered is a function of the probability before times D ("Divine Indicator Scale"): 10 indicates the evidence is 10 times more likely to be produced if God exists, 2 is two times more likely if God exists, 1 is neutral, .5 is moderately more likely if God does not exist, and 0.1 is much more likely if God does not exist. Unwin offers the following figures for six lines of evidence: *recognition of goodness* (D=10), *existence of moral evil* (D=0.5), *existence of natural evil* (D=0.1), *intranatural miracles* (prayers) (D=2), *extranatural miracles* (resurrection) (D=1), and *religious experiences* (D=2).

Plugging these figures into the above formula (in sequence, where the P_{after} figure for the first computation is used for the P_{before} figure in the second computation, and so on for all six Ds), Unwin concludes,

"The probability that God exists is 67 percent." Remarkably, Unwin then confesses, "this number has a subjective element since it reflects *my* assessment of the evidence. It isn't as if we have calculated the value of pi for the first time."

Indeed, based on my own theory of the evolutionary origins of morality and the sociocultural foundation of God beliefs and religion, starting with a 50 percent probability of God's existence (as Unwin does) I estimate these probabilities: recognition of goodness (D=0.5), existence of moral evil (D=0.1), existence of natural evil (D=0.1), intranatural miracles (D=1), extranatural miracles (D=0.5), and religious experiences (D=0.1). With these figures, the probability that God exists is 0.02, or 2 percent.

Regardless, this subjective component in the formula relegates the process to an entertaining exercise in thinking—on a par with mathematical puzzles—but little more. In my opinion, the God question is a scientifically insoluble one. Thus all such scientistic theologies are compelling only to those who already believe. Religious faith depends on a host of social, psychological, and emotional factors that have little or nothing to do with probabilities, evidence, and logic. This is faith's inescapable weakness. It is also, undeniably, its greatest power.

72

Miracle on Probability Street

The Law of Large Numbers guarantees that one-in-a-million miracles happen 321 times a day in America

Because I am often introduced as a "professional skeptic," people feel compelled to challenge me with stories about highly improbable events. The implication is that if I cannot offer a satisfactory natural explanation for *that particular event*, the general principle of supernaturalism is preserved. A common story is the one about having a dream or thought about the death of a friend or relative, then receiving a phone call five minutes later about the unexpected death of that very person.

Although I cannot always explain such specific occurrences, a principle of probability called the Law of Large Numbers shows that an event with a low probability of occurrence in a small number of trials has a high probability of occurrence in a large number of trials. In other words, million-to-one odds happen 321 times a day in America.

In their delightful 2004 book *Debunked!* (Johns Hopkins University Press), CERN physicist Georges Charpak and University of Nice physicist Henri Broch show how the application of probability theory to such events is enlightening. In the case of death premonitions, suppose you know of ten people a year who die, and that you think about each of those people once a year. One year contains 105,120 five-minute intervals during which you might think about each of the ten people, a probability of

1 out of 10,512, certainly an improbable event. However, there are 321 million Americans (in 2015). Assuming, for the sake of our calculation, that they think like you, 1/10,512 x 321,000,000 = 30,537 people per year, or 84 people per day for whom this improbable premonition becomes probable. With the *confirmation bias* firmly in force (where we notice the hits and ignore the misses in support of our favorite beliefs), if just a couple of these people recount their miraculous tales in a public forum (next on *Oprah*!), the paranormal seems vindicated. In fact, it is nothing more than the laws of probability writ large.

Another form of this principle was suggested by Institute for Advanced Study physicist Freeman Dyson in a review of *Debunked!* (*New York Review of Books,* March 25, 2004), as "Littlewood's Law of Miracles" (John Littlewood was a Cambridge University mathematician): "In the course of any normal person's life, miracles happen at a rate of roughly one per month." Dyson explains: "During the time that we are awake and actively engaged in living our lives, roughly for eight hours each day, we see and hear things happening at a rate of about one per second. So the total number of events that happen to us is about thirty thousand per day, or about a million per month. With few exceptions, these events are not miracles because they are insignificant. The chance of a miracle is about one per million events. Therefore we should expect about one miracle to happen, on the average, every month."

Despite this cogent explanation, Dyson concludes with a "tenable" hypothesis that "paranormal phenomena may really exist" because, he says, "I am not a reductionist" and "that paranormal phenomena are real but lie outside the limits of science is supported by a great mass of evidence." That evidence is entirely anecdotal, he admits, but because his grandmother was a faith healer and his cousin edits the *Journal of Psychical Review,* and because anecdotes gathered by the Society for Psychical Research and other organizations suggest that under certain conditions (e.g., stress), some people sometimes exhibit some paranormal powers (unless experimental controls are employed, at which point the powers disappear), "I find it plausible that a world of mental phenomena should exist, too fluid and evanescent to be grasped with the cumbersome tools of science."

Freeman Dyson is one of the great minds of our time and I admire him immensely. But even genius of this magnitude cannot override the cognitive biases that favor anecdotal thinking. The only way to find out if anecdotes represent real phenomena is controlled tests. Either people can read other people's minds (or ESP cards) or they can't. Science has unequivocally demonstrated that they can't. QED. And being a holist instead of a reductionist, being related to psychics, or reading about weird things that befall people does not change this fact.

73

Mustangs and Monists

The dualist belief that body and soul are separate entities is natural, intuitive, and with us from infancy. It is also very probably wrong

When I was seventeen I purchased my dream car—a 1966 Ford Mustang, blue with a white vinyl roof, bucket seats, and a powerful eight-cylinder 289-cubic-inch engine that could peg a speedometer at 140 mph. As testosterone overloaded young men are wont to do, however, over the course of the next fifteen years I systematically wrecked and replaced nearly every part of that car, such that by the time I sold it in 1986 there was hardly an original part remaining. Nevertheless, I turned a tidy profit because my "1966" Mustang was now a collector's classic. Even though the physical parts were not original, the essence of its being—its "Mustangness"—was that model's pattern. My Mustang's essence—its "soul"—is more than a pile of parts; it is a pattern of information arranged in a particular way.

The analogy applies to humans and souls. The actual atoms and molecules that make up my brain and body today are not the same ones as when I was born on September 8, 1954, more than sixty years ago. Nevertheless, I am still "Michael Shermer," the pattern of information coded in my DNA and neural memories. My friends and family do not treat me any differently from moment to moment, even though atoms and molecules are cycling in and out of my body and brain, because they assume that the basic pattern remains unchanged. My soul is a pattern of information.

Dualists hold that body and soul are separate entities and that the soul will continue beyond the existence of the physical body. Monists contend that body and soul are the same and that the death of the body—the disintegration of DNA and neurons that store the pattern of information—spells the end of the soul. Until a technology is developed to download our patterns into a more durable medium than the electric meat of our carbon-based protein (silicon chips is one suggestion), when we die our patterns die with us.

The principal barrier to a general acceptance of the monist position is that it is counterintuitive. As Yale University psychologist Paul Bloom argues in his intriguing book *Descartes' Baby* (Basic Books, 2004), we are natural-born dualists. Children and adults alike speak of "my body" as if "my" and "body" are dissimilar. In one among many experiments Bloom recounts, for example, young children are told a story about a mouse that gets munched by an alligator. The children agree that the mouse's body is dead—it does not need to go to the bathroom, it can't hear, and its brain no longer works. However, they insisted that the mouse is still hungry, concerned about the alligator, and wants to go home. "This is the foundation for the more articulated view of the afterlife you usually find in older children and adults," Bloom explains. "Once children learn that the brain is involved in thinking, they don't take it as showing that the brain is the source of mental life; they don't become materialists. Rather, they interpret 'thinking' in a narrow sense, and conclude that the brain is a cognitive prosthesis, something added to the soul to enhance its computing power."

The reason dualism is intuitive is that the brain does not perceive itself and so imputes mental activity to a separate source. Hallucinations of preternatural beings (ghosts, angels, aliens) are perceived as real entities, out-of-body and near-death experiences are experienced as external events, and the pattern of information that is our memories, personality, and "self" is sensed as a soul.

Is scientific monism in conflict with religious dualism? Yes it is. Either the soul survives death or it does not, and there is no scientific evidence that it does. Does monism extirpate all meaning in life? I think not. If this is all there is, then every moment, every relationship, and every person count, and count more if there is no tomorrow than if there is, for it elevates all of us to a higher plane of humanity and humility that we are in this limited time and space together, a momentary proscenium in the drama of the cosmos.

74

Flying Carpets and Scientific Prayer

Scientific experiments claiming that distant intercessory prayer produces salubrious effects are deeply flawed

In late 1944, as he cajoled his flagging troops to defeat the Germans in the Battle of the Bulge, General George S. Patton turned to the chief chaplain of his Third Army, James H. O'Neill, for help:

> PATTON: Chaplain, I want you to publish a prayer for good weather. I'm tired of these soldiers having to fight mud and floods as well as Germans. See if we can't get God to work on our side.
>
> O'NEILL: Sir, it's going to take a pretty thick rug for that kind of praying.
>
> PATTON: I don't care if it takes the flying carpet. I want the praying done.

Although few attribute Patton's subsequent success to a divine miracle, in recent years a number of papers have been published in peer-reviewed scientific journals claiming that distant intercessory prayer leads to health and healing. These studies are fraught with methodological problems.

1. *Fraud.* In 2001, the *Journal of Reproductive Medicine* published a study by three Columbia University researchers claiming that prayer for women undergoing in-vitro fertilization resulted in a pregnancy rate of 50 percent,

double that of women who did not receive prayer. Media coverage was extensive. ABC News medical correspondent Dr. Timothy Johnson, for example, reported, "A new study on the power of prayer over pregnancy reports surprising results; but many physicians remain skeptical." One of those skeptics was a University of California clinical professor of gynecology and obstetrics named Bruce Flamm, who not only found numerous methodological errors in the experiment, but also discovered that one of the study's authors, Daniel Wirth (aka "John Wayne Truelove"), is not an MD but an MS in parapsychology who has since been indicted on felony charges for mail fraud and theft, for which he pled guilty. The other two authors have refused comment, and after three years of inquiries from Flamm the journal removed the study from its web site and Columbia University launched an investigation.

2. *Lack of controls.* Many of these studies failed to control for such intervening variables as age, sex, education, ethnicity, socioeconomic status, marital standing, degree of religiosity, and the fact that most religions have sanctions against such insalubrious behaviors as sexual promiscuity, alcohol and drug abuse, and smoking. When such variables are controlled for, the formerly significant results disappear. One study on recovery from hip surgery in elderly women failed to control for age; another study on church attendance and illness recovery did not consider that people in poorer health are less likely to attend church; a related study failed to control for levels of exercise.

3. *Outcome differences.* In one of the most highly publicized studies of cardiac patients prayed for by born-again Christians, twenty-nine outcome variables were measured but on only six did the prayed-for group show improvement. In related studies, different outcome measures were significant. To be meaningful, the same measures need to be significant across studies, because if enough outcomes are measured some will show significant correlations by chance.

4. *File-drawer problem.* In several studies on the relationship between religiosity and mortality (religious people allegedly live longer), a number of religious variables were used, but only those with significant

correlations were reported. Meanwhile, other studies using the same religiosity variables found different correlations and, of course, only reported those. The rest were filed away in the drawer of nonsignificant findings. When all variables are factored in together, religiosity and mortality show no relationship.

5. *Operational definitions.* When experimenting on the effects of prayer, what, precisely, is being studied? For example, what type of prayer is being employed? (Are Christian, Jewish, Muslim, Buddhist, Wiccan, and shamanistic prayers equal?) Who or what is being prayed to? (Are God, Jesus, and a universal life force equivalent?) What are the length and frequency of the prayer? (Are two ten-minute prayers equal to one twenty-minute prayer?) How many people are praying, and does their status in the religion matter? (Is one priestly prayer identical to ten parishioner prayers?) Most prayer studies either lack such operational definitions, or there is no consistency across studies in such definitions.

The ultimate fallacy is theological: if God is omniscient and omnipotent, He should not need to be reminded or inveigled that someone needs healing. Scientific prayer makes God a celestial lab rat, leading to bad science and worse religion.

75

Bowling for God

Is religion good for society? Science's definitive answer: it depends

Are religion and belief in God necessary components of social health? The data are conflicting.

On the one hand, in a 2005 study published in the *Journal of Religion and Society*—"Cross-National Correlations of Quantifiable Societal Health with Popular Religiosity and Secularism in the Prosperous Democracies"—independent scholar Gregory S. Paul found an inverse correlation between religiosity (measured by belief in God, biblical literalism, and frequency of prayer and service attendance) and societal health (measured by rates of homicide, suicide, childhood mortality, life expectancy, sexually transmitted diseases, abortion, and teen pregnancy) in eighteen developed democracies. "In general, higher rates of belief in and worship of a creator correlate with higher rates of homicide, juvenile and early adult mortality, STD infection rates, teen pregnancy, and abortion in the prosperous democracies," Paul found. "The United States is almost always the most dysfunctional of the developed democracies, sometimes spectacularly so." Indeed, the United States scores the highest in religiosity and the highest (by far) in homicides, STDs, abortions, and teen pregnancies. So despite being the most religious nation of the sample (not to mention the most economically prosperous), the United States is at or near the bottom of every societal health measure.

On the other hand, Syracuse University professor Arthur C. Brooks, in *Who Really Cares* (Basic Books, 2006), argues that when it comes to charitable giving and volunteering, numerous quantitative measures debunk the myth of "bleeding heart liberals" and "heartless conservatives." Conservatives donate 30 percent more money than liberals (even when controlled for income), give more blood, and log more volunteer hours. In general, religious people are four times more generous than secularists to all charities, 10 percent more munificent to nonreligious charities, and 57 percent more likely than a secularist to help a homeless person. Those raised in intact and religious families are more charitable than those who are not. In terms of societal health, charitable givers are 43 percent more likely to say they are "very happy" than nongivers, and 25 percent more likely than nongivers to say their health is "excellent" or "very good." As well, the working poor give a substantially higher percentage of their incomes to charity than any other income group, and three times more than those on public assistance of comparable income—poverty is not a barrier to charity, but welfare is. One explanation for these findings is that people who are skeptical of big government give more than those who believe that the government should take care of the poor. "For many people," Brooks explains, "the desire to donate other people's money displaces the act of giving one's own."

However, correlation is not causation, left and right are not so religiously cleaved, and comparing data from between nations and from within nations is problematic. Nevertheless, according to Harvard University professor Pippa Norris and University of Michigan professor Ronald Inglehart, in their book *Sacred and Secular* (Cambridge University Press, 2004), data from the Comparative Study of Electoral Systems analyzing thirty-seven presidential and parliamentary elections in thirty-two nations since 1990 found that 70 percent of the devout (attend religious services at least once per week) voted for parties of the right, compared with only 45 percent of the secular (never attend religious services). The effect is striking in America. In the 2000 US presidential election, for example, "religion was by far the strongest predictor of who voted for Bush and who voted for Gore—dwarfing the explanatory power of social class, occupation, or region."

The theory of "social capital" may help resolve these disparate findings. As defined by Robert Putnam in his book *Bowling Alone* (Simon & Schuster, 2000), social capital means "connections among individuals—social networks and the norms of reciprocity and trustworthiness that arise from them." In their analysis of data from the World Values Survey, for example, Norris and Inglehart found a positive correlation between "religious participation" and membership in "non-religious community associations," including women's, youth, peace, social welfare, human rights, and environmental conservation groups (and, apparently, bowling leagues). "This pattern confirms social capital theory's claim that the social networks and personal communications derived from regular churchgoing play an important role, not just in promoting activism within religious-related organizations, but also in strengthening community associations more generally. By providing community meeting places, linking neighbors together, and fostering altruism, in many (but not all) faiths, religious institutions seem to bolster the ties of belonging to civic life."

Religious social capital leads to charitable generosity and group membership but does comparatively worse than secular social capital for such ills as homicides, STDs, abortions, and teen pregnancies. Three reasons suggest themselves: (1) these problems have other causes entirely; (2) secular social capital works better for such problems; and (3) these problems are related to what we might call moral capital, or the connections within an individual between morality and behavior that are best fostered within families, the fundamental social unit in our evolutionary history that arose long before religions and governments. Thus moral restraints on aggressive and sexual behavior are best reinforced by the family, be it secular or sacred.

ACKNOWLEDGMENTS

Thank you Serena Jones, Allison Adler, Rita Quintas, Carolyn O'Keefe, Maggie Richards, and Paul Golob at my publisher Henry Holt for bringing this book to fruition, and to Katinka Matson, John Brockman, Max Brockman, Russell Weinberger, and the staff of Brockman, Inc., my literary agency.

Thanks also go to my colleagues and associates at the Skeptics Society and *Skeptic* magazine, including Nickole McCullough, Ann Edwards, Daniel Loxton, William Bull, Jerry Friedman, and, most especially, my partner Pat Linse. Our many volunteers who make such an organization run smoothly deserve acknowledgment: senior editor Frank Miele; senior scientists David Naiditch, Bernard Leikind, Liam McDaid, Claudio Maccone, Thomas McDonough, and Donald Prothero; contributing editors Tim Callahan, Harriet Hall, and Carol Tavris; editors Sara Meric and Kathy Moyd; photographer David Patton and videographer Brad Davies; and our many volunteers: Jaime Botero, Bonnie Callahan, Tim Callahan, Cliff Caplan, Michael Gilmore, Diane Knutdson, and Teresa Lavelle. Thanks as well for the institutional support for the Skeptics Society at the California Institute of Technology go to Cindy Dale, Dwayne Miles, Hall Daily, and Laurel Auchampaugh. As well, I am in debt to my lecture agent (who is now also my friend), Scott Wolfman, and his team at Wolfman Productions (Diane Thompson and Miriam Pachniuk) for their contributions in bringing science and skepticism to the speakers' circuit.

Of all the writing I do none means more to me than my monthly column in *Scientific American*, which I began in April 2001. My editor, Mariette DiChristina, is in fact the editor in chief and the first woman to hold that august post in this, the longest continuously published magazine in American history (175 years and counting). To her, my editor Fred Guterl, and especially my previous editor who gave me my break, John Rennie, I owe a deep debt of gratitude.

Finally, to my family I acknowledge their love and support: my daughter Devin Shermer, my wife Jennifer Shermer, and my sister Tina Shermer, to whom this book is dedicated. I've got your back.

INDEX

Page numbers in *italics* refer to illustrations.

ABOUT THE AUTHOR

Dr. Shermer received his BA in psychology from Pepperdine University (1976), his MA in experimental psychology from California State University, Fullerton (1978), and his PhD in the history of science from Claremont Graduate University (1991). He is a Presidential Fellow at Chapman University. He has been a college professor since 1979, teaching psychology, evolution, and the history of science at Occidental College and Glendale College. As a public intellectual he regularly contributes opinion editorials, book reviews, and essays to the *Wall Street Journal*, the *Los Angeles Times*, *Science*, *Nature*, and other publications. He has appeared on such shows as the *Colbert Report*, *20/20*, *Dateline*, *Charlie Rose*, *Larry King Live*, *Oprah*, *Unsolved Mysteries*, and other shows, as well as interviews in countless science and history documentaries aired on PBS, A&E, Discovery, the History Channel, the Science Channel, and the Learning Channel. Dr. Shermer was the cohost and coproducer of the thirteen-hour Family Channel television series *Exploring the Unknown*. He is the founding publisher of *Skeptic* magazine, the executive director of the Skeptics Society (www.skeptic.com), and a monthly columnist for *Scientific American*.